Thin Safety Margin

Thin Safety Margin

The SEFOR Super-Prompt-Critical Transient Experiments

Ozark Mountains, Arkansas, 1970–71

Jerry Havens
Collis Geren

ARKANSAS SCHOLARLY EDITIONS
FAYETTEVILLE
2021

Manufactured in the United States of America

ISBN: 978-1-68226-174-3
eISBN: 978-1-61075-749-2

25 24 23 22 21   5 4 3 2 1

∞ The paper used in this publication meets the minimum requirements of the American National Standard for Permanence of Paper for Printed Library Materials Z39.48-1984.

Library of Congress Control Number: 2021941873

Preface

This book began as a history of the Southwest Experimental Fast Oxide Reactor (SEFOR). Our research led us to additional goals—to inform the public generally about the risks attending a national commitment to utilize fast nuclear reactors for electric power generation, and perhaps most importantly to us in the end, to consider carefully the risks of a worst-case accident that were taken during the experiments conducted in the SEFOR 20 megawatt (thermal) plutonium-fueled fast-neutron reactor during the period 1969–71 in the rural Ozark Mountains of Arkansas. The book is dedicated to:

Richard E. Webb, 1939–
and
David Okrent, 1922–2012

Whom we consider pioneers in fast-reactor safety engineering.

Drs. Webb and Okrent, practicing nuclear engineers with extensive experience in fast-reactor safety research and development, were deeply involved in the debates regarding fast-reactor safety during the period 1960–80 when major commitments to a liquid metal fast breeder reactor (LMFBR) based electric power production program were high-priority goals for the U.S. Atomic Energy Commission (AEC). We considered the AEC's response to Webb's and Okrent's officially documented advice regarding the risks of accidental explosions in fast reactors that could fail the containment structures provided. It appears, as the United States considers proposals to initiate a fast-reactor-based program for electric power production as a means of dealing with the climate change threat, that there are critically important lessons here in catastrophic risk management as well as in government.

Contents

Prologue

"If something seems too good to be true, it probably is."

During the 1960s, there arose a promise of electric power production so cheap that it might not even have to be metered. And it arose as the result of the United States entering the nuclear age via the atomic bomb destruction of Hiroshima and Nagasaki. In the ultimate swords-to-plowshares project, it was theorized that it is possible to design a nuclear reactor that actually produces more nuclear fuel than it consumes in the process of generating heat to boil water to provide steam to drive large generators of electricity. Ignoring the problem of residual radioactive waste generated by the operation of any nuclear reactor, this proposed reactor would have to be a "fast" reactor in order to actually produce more fuel than used. The fuel produced by these "breeder reactors" would be plutonium. A number of individuals raised concerns about the advisability of developing fast reactors in general, the concern being the safety of such reactors. But nuclear physics had predicted a self-regulation mechanism inherent in such reactors called the Doppler effect that could possibly remove concerns about safety.

At this time in American history, the Atomic Energy Commission (AEC) was in control of the nuclear programs in the United States. The AEC was unique in that it not only regulated nuclear activities, it also promoted those activities, and it had a desire to experimentally test the effectiveness of the Doppler effect in an effort to promote breeder reactor development. One would suppose a national laboratory would be charged by the federal government with designing such an experiment. However, this was not the case. A consortium of energy-producing industries and foreign governments called the Southwest Experimental Fast Oxide Reactor (SEFOR) Consortium officially initiated the construction of a 20-megawatt reactor to test the Doppler effect, and that reactor would be sited within twenty miles south and west of

Fayetteville, Arkansas, the home of the University of Arkansas. The AEC provided the nuclear fuel for this reactor. That fuel was plutonium, and the amount was sufficient to build some one hundred atomic bombs of the size of those used on Japan if one desired to do so. The actual design and operation of the reactor and the actual experimental procedures were accomplished by General Electric.

The experiments were simple: have the reactor operating in a manner capable of generating enough heat to generate up to 20 megawatts of thermal power; then remove the control mechanism that maintained that level of operation and see if the Doppler effect would actually slow the now out-of-control reactor. General Electric did an excellent job of designing the reactor and process as the reactor was brought back under control in a series of experiments, each exceeding the previous as to the level of "out-of-control" that occurred. In the last test, conducted with the reactor operating at 8 megawatts power, the amount of heat produced rose to nearly 10,000 megawatts, an increase by a factor of more than a thousand, before the reactor was brought back under control.

As this book shows, this SEFOR experiment was hailed by the nuclear industry as one of several demonstrations that fast reactors could be controlled. The project was terminated at the end of that month (December 1971), and the site closed in early 1972. In 1975, the SEFOR Consortium donated the SEFOR site to the University of Arkansas.

The question this book explores is: just exactly what did SEFOR prove? Since there was no runaway nuclear reaction, the experiment was a success, but just how controlled was the actual experiment? To further raise questions, remember that the fuel was plutonium, which causes radiation poisoning and persists for thousands of years. Also consider the amount present in the small SEFOR reactor. And finally, consider that fast reactors require a more concentrated form of nuclear fuel (approaching weapons grade) than the less concentrated fuels in today's nuclear reactors used for electric power generation.

In the global warming concerns of today, one again hears that nuclear should be pushed to the forefront, and breeder reactors are the way to go. This book provides details of exactly what SEFOR accomplished and why those in positions of authority should very carefully consider exactly what we know about the safety of fast reactors.

The authors were faculty at the University of Arkansas for a combined total of more than eighty years of service. Both were actively involved in the process of obtaining the data on the creation and function of SEFOR in preparation for seeking federal funding in totally dismantling the site. Our first approach to federal funding was rebuffed because the federal government claimed no function in the creation of SEFOR. Thanks to many years of Arkansas legislators working for this goal, federal funding was finally obtained, and the site is now a greenfield available for general use.

It is of interest to note that the AEC was abolished in 1974 due to congressional concerns about its ability to regulate nuclear activities.

Thin Safety Margin

CHAPTER

1

Introduction

There is an understandable drive on the part of men of good will to build up the positive aspects of nuclear energy simply because the negative aspects are so distressing.

Alvin M. Weinberg, 1915–2006

Until late 2018, the remains of the SEFOR reactor rested in a 114.5-feet-high, 50-feet-diameter, steel cylinder intended as a final barrier to prevent catastrophic release of radioactive materials to the atmosphere if the reactor inside were ever breached by explosion. The top half of the rusting containment vessel protruded from a hayfield in the rural Ozark Mountains nineteen miles south-southwest of Fayetteville, Arkansas, the home of the University of Arkansas.

SEFOR's construction was completed in the late 1960s, and the research reactor was operated by the General Electric Company for the U.S. Atomic Energy Commission for about three years. The authors, now retired professors at the university, were involved in one way or another with the reactor site for most of our careers. Our involvement enabled our access to the reactor's history; we believe that history is unique in the development of the nuclear power generation industry.

SEFOR was closed in early 1972 following a multiyear reactor-safety research program. The idle reactor site was deeded in 1975 to the university by the Southwest Atomic Energy Associates (SAEA), a consortium of electric power utility companies in the area. The SAEA's promotional literature implied that completion of a large array of fast-breeder reactors "burning" plutonium fuel would ensure production of electric power almost "too cheap to meter" and described SEFOR as "the most significant single reactor-safety experiment in the western world." In recognition of the importance of the reactor-safety research completed at the site, the American Nuclear Society in 1986 designated SEFOR a Nuclear Historic Landmark. In the Mechanical Engineering Building at the university, which earlier housed its nuclear engineering program, a plaque read:

> SEFOR
> Resolved a key LMFBR safety issue by demonstrating the inherent negative prompt-Doppler power coefficient in mixed plutonium-uranium oxide fuel.

As this book went to press, forty-eight years after the reactor was closed and placed in temporary "SAFSTOR" condition, the University of Arkansas had received congressional authorization, backed by funding of approximately $28 million, to complete a decommissioning process that would return the site to "greenfield" condition. By late July 2019, the reactor vessel had been removed from its containment, placed in a special protective vessel, and trucked to a Nevada disposal site.

But already by the time SEFOR commenced operation a potentially serious *nuclear* explosion risk had been identified, and it appears now that the principal purpose for SEFOR was to demonstrate evidence of a theoretically predicted safety factor that could provide an answer for that concern. Up to this time (around 1960), commercial nuclear electric-power reactors, operating with fissile fuel concentrations of less than about 3%, had been accepted as having an inherent safety characteristic—there appeared to be no way the position of the fuel could be rearranged by accident that could cause a nuclear-bomb-like explosion. SEFOR, in contrast, required nearly a tenfold enrichment of its plutonium fuel, and such enrichment had been predicted to enable fission reaction rates that could result in nuclear explosions powerful enough to destroy the reactor-containment systems. Dealing with this potentially dire safety problem remains today a critically contentious proposition.

Almost half a century later, there are renewed calls for a fast-neutron fission reactor program for electric power generation; proponents claiming that fast-neutron reactors are the best solution available to meet the country's electric power energy needs while simultaneously reducing carbon dioxide additions (from burning fossil fuels) to the atmosphere. However, this book shows that very serious additional concerns remain about the hazards attending operation of fast-fission nuclear reactors, even now when we are not confident of our ability to "engineer" our current aging fleet of water-cooled (thermal) nuclear power reactors to provide satisfactory safety to the public in the event of "highly unlikely" events such as are history now (several times over).

We believe this book presents a convincing argument that the SEFOR sodium-cooled-enriched-plutonium reactor was an important step in the development of reactors driven by fast neutrons capable of reaching, in accident conditions, fission rates approaching nuclear bomb capability. While the nuclear explosions that became possible in such reactors would be extremely "inefficient" compared to a well-designed bomb, their potential to catastrophically rupture any containment structure that could be economically provided could be so high as to be deemed completely unacceptable. Indeed, we do not take lightly our belief that such an explosion as could rupture its containment structure and release a large fraction of plutonium into the environment is a very real example of the "dirty (radioactive) bomb" fear that worries authorities so seriously.

SEFOR successfully demonstrated the inherent (Doppler) safety effect in a fast reactor using enriched plutonium-oxide fuel. The reactor site was closed in early 1972, and the AEC stepped up research and development programs to support building a large fleet (roughly a thousand were planned) of liquid-metal-cooled, plutonium-oxide-fueled, fast-breeder reactors (LMFBRs) to meet the energy needs of the country projected for the year 2000. But those plans received much less attention following the landmark decision in 1983 to abandon the Clinch River (Tennessee) Breeder Reactor (CRBR) Demonstration Plant. The reasons for the CRBR's cancellation continue to be debated, but there is little doubt that the cancellation resulted at least partly from two developments: a less pessimistic outlook for fissile uranium availability and remaining questions of public safety associated with the heightened nuclear explosion risk. It seemed clear that the uncertainties in cost of ensuring public safety by building fast-reactor containment structures sufficiently strong to confidently prevent catastrophic accidental releases of radioactive materials to the environment importantly worsened the economic outlook.

SEFOR was closed during the period when the LMFBR program had achieved highest priority with the AEC. But the sufficiency of the SEFOR-demonstrated Doppler effect to ensure prevention of runaway nuclear explosions in fast reactors was still being questioned by experts. The importance of such questions had been highlighted by accidents resulting in a partial meltdown in the EBR-I fast-breeder experimental reactor in 1955 and a partial meltdown/explosion in the SL-1 fast reactor in 1961, both at the AEC's National Reactor Test Station in Idaho, and in the partial meltdown in 1966 of the FERMI-1 fast-breeder demonstration reactor constructed by Detroit Edison on Lake Erie about twenty miles from Detroit. The accidents at SL-1 and FERMI-1 have been described for the lay reader in the books *Idaho Falls*[1] and *We Almost Lost Detroit*,[2] respectively

Today, despite claims that SEFOR provided a "positive" answer to the nuclear-explosion-critical-safety question, it appears that the possibility of an uncontrolled nuclear explosion that could fail any containment that could be practicably and economically provided still exists. At the same time, many experts contend that the consequences of even a very weak nuclear explosion in a large fast reactor sufficiently powerful to vaporize and release a substantial fraction of its fissile fuel along with the radioactive components of the spent fuel to the atmosphere would be so severe as to be deemed a completely unacceptable risk.

Much has been written about the AEC's structuring of the nation's nuclear weapons program and its connection with the nuclear power industry. Our interest was sharply piqued in such matters following the transfer of ownership of the SEFOR site to the University of Arkansas in 1975. By that time, SEFOR was becoming viewed as a potentially hazardous nuclear/chemical waste site, and the university learned that the federal government was not disposed to accept any responsibility for the government-required cleanup operations that were mandated. Then our research on the history of the SEFOR project produced, in Appendix G of Thomas Cochran's *The Liquid Metal Fast Breeder Reactor: An Environmental and Economic Critique*,[3] excerpts from testimony in 1972 of Dr. Richard E. Webb before the Joint Committee on

Atomic Energy of the U.S. Congress[4] regarding the explosion potential of fast reactors. Dr. Webb received his PhD in nuclear engineering from Ohio State University in 1972 with his dissertation entitled "Some Autocatalytic Effects during Explosive Power Transients in Liquid Metal Cooled Fast Breeder Nuclear Power Reactors (LMFBRs)."[5] Dr. Webb testified (in part) that "With one-half ton of Plutonium in the SEFOR reactor, it appears that the AEC simply took a chance with the public safety by purposely causing power excursions, which one tries normally to prevent in power reactors, to test a safety effect (Doppler feedback) that was not beforehand demonstrated in a fast-reactor power excursion."

This book focuses on Dr. Webb's documented statements about SEFOR along with other material that he submitted to the AEC on the subject of "runaway" nuclear explosion risks attending the operation of fast reactors. We have also relied on Dr. Webb's book *The Accident Hazards of Nuclear Power Plants,*[6] published by the University of Massachusetts Press in 1976. It appears that the AEC failed to give serious consideration to Dr. Webb's science-based advice to Congress's Joint Committee on Atomic Energy on the explosion potential attending operation of fast-breeder reactors.

Our research revealed the international significance of the safety experiments conducted quietly in the rural Ozark Mountains of northwest Arkansas. The SEFOR research project resulted from international agreements, backed by financial assistance, among the United States, Germany, Holland, Belgium, and Luxembourg. A series of difficulties beset the project due to U.S. laws that prohibited the authorization of atomic projects in the United States that could be controlled or dominated by a foreign power, and the details of resolution of such disputes clearly indicated the power that could be brought to bear to ensure continuation of the SEFOR project. SEFOR was highly important to the AEC and the representatives of government organizations then promoting nuclear power in Europe, particularly Germany.

The AEC was abolished in 1974. The United States' Clinch River Breeder Reactor (CRBR) was canceled, uncompleted, in 1983. Germany's SNR-300 breeder reactor (Germany's demonstration breeder reactor) was completed and fueled for operation in 1985. Never operated, the SNR-300 was officially canceled in 1991. Both the CRBR and the

Note to Readers

A primary goal of this book began as an appeal to three very different groups of people to think critically about a major risk to which they are considering a commitment. That commitment is the adoption of fast nuclear reactor technology for generating electric power—buttressed with the argument that the "nuclear" approach is the best means of addressing the climate change threat of increasing carbon dioxide levels in the atmosphere. The groups of people are (a) experts, (b) students, and (c) the general public. We suspect it's unlikely that those professing expertise in fast-neutron nuclear electric power generation as well as climate change will all be convinced by the arguments presented, but we are hopeful of their careful consideration. While the public, if only due to their number, will ultimately decide these questions if democratic choices prevail, a special obligation was felt to address the university student population. It appears that students of all ages have insufficient knowledge to decide such issues on their own, or perhaps more importantly, to meaningfully influence the general public in that regard. We received a barrage of warnings that we had little chance of reaching all three such disparate groups—the concern was that some of the material presented was simply too technical for all but those expert in nuclear science. We have tried to present a reasonably clear picture of the risks attending the use of fast-neutron reactors for electric power generation with a minimum of nuclear science theory. Where this was simply not practical, we have collected the required "technical" material largely in chapter 5. For those mathematically less inclined and hoping primarily to equip themselves with important information to help decide the issue of risk attending fast-neutron fission generation of electrical power, the heavy sledding involved in chapter 5, although critically important to the expert community, is not absolutely required. But please, if you do skip chapter 5, do not miss the facts presented therein for the record by Dr. Richard Webb to the AEC regarding the risks taken in the SEFOR experiments. We believe Webb's admonitions to the AEC about the risks associated with LMFBR operations were never nearly sufficiently considered. We hope that the information presented here, some of which is not easy to obtain from the popular literature, has been reduced to a level simple and straightforward enough to be appreciated by all three groups and that all will find it useful.

Notes

1. William McKeown *Idaho Falls: The Untold Story of America's First Nuclear Accident,* ECW Press, 2003.

2. John G. Fuller, *We Almost Lost Detroit,* Reader's Digest Press, 1975.

3. Thomas B. Cochran, *The Liquid Metal Fast Breeder Reactor: An Environmental and Economic Critique*, Resources for the Future, distributed by Johns Hopkins Press.

4. "Liquid Metal Fast Breeder Reactor (LMFBR) Demonstration Plant," Hearings before the Joint Committee on Atomic Energy of the U.S. Congress, 92 Congress (Sept. 7, 8, 12, 1972), pp, 179–187.

5. Richard E. Webb, "Some Autocatalytic Effects during Explosive Power Transients in Liquid Metal Cooled, Fast Breeder Nuclear Power Reactors (LMFBRs)" (PhD diss., Ohio State University, 1971).

6. Richard E. Webb, *The Accident Hazards of Nuclear Power Plants*, University of Massachusetts Press, 1976.

CHAPTER 2

SEFOR Site, Strickler, Arkansas, December 1971

After almost two years of preparatory experiments, the General Electric Company was confident that the theory underlying the reactor's design was correct and the experiments beginning in December would demonstrate that SEFOR, fueled with plutonium oxide instead of pure metallic plutonium used in some fast reactors (and in atomic bombs), would respond to a *planned /intentional* upset (reactivity increase) condition with a sufficient time lag that would allow the reactor's energy output to be slowed and stopped without suffering an explosion that could endanger the reactor's containment structure.

The *demonstration* of the "inherent Doppler safety" effect in a fast plutonium-oxide-fuel reactor that was *super-prompt-critical* (chain-fission-reacting at nuclear bomb rates) *was* the principal purpose of SEFOR's construction and operation. The company's reactor designers appear to have planned carefully. In addition to their confidence that the Doppler-theory-predicted time lag would be observed, the experiment was designed so that even if the anticipated (Doppler) time lag were not of the magnitude anticipated by theory predictions, the potential for the reaction to heat the fuel sufficiently to melt in amounts sufficient to cause severe explosion damage was not thought to be there anyway, obviating an explosion that could not be contained.

It was extremely important to be sure that no detail had been overlooked. The reactor fuel comprised approximately 900 pounds of plutonium, a quantity sufficient in purified form to build approximately one hundred atomic bombs each with the explosive power of the bombs dropped on Hiroshima and Nagasaki ending World War II. More importantly to Arkansas and the adjacent states of Oklahoma, Kansas, and Missouri, if it were possible for an explosion to vaporize a substantial amount of the plutonium in the reactor, and that plutonium escaped the containment into the atmosphere in the form of an aerosol, the surroundings extending to distances of many miles could be catastrophically affected, even forcibly abandoned.

Six experiments were completed during the coming week; in each succeeding experiment, an increased amount of positive reactivity was introduced into the reactor's core. The amount of reactivity inserted increased from amounts sufficient to produce sub-prompt criticality *to, and slightly beyond.* the amount that enabled *super-prompt-criticality,* the condition required to achieve reaction speeds with nuclear bomb explosion potential.

The experiments appear to have been planned to introduce in each successive experiment an additional amount of reactivity that was sufficiently small to be canceled by the predicted Doppler effect, thus preventing the melting temperature of the plutonium-oxide fuel from being reached.

The ramp reactivity insertion in each of the six experiments was intentionally short-lived, intended to be limited to 0.1-second duration. Following the reactivity insertion, each experiment was terminated with a time-delayed negative-reactivity insertion (reactor SCRAM) initiated approximately 0.35 seconds after the transient reactivity insertion. The SCRAM delay time following the reactivity insertion was planned to allow time for the *demonstration* of the Doppler effect to cause the extremely rapidly rising reactor power rate to be halted and proceed downward to a level where the reactor could be SCRAMMED.

The entire "business" of each of the six final experiments of the program, including the super-prompt-critical experiments, was thus completed in less than one second. In the last test, No. 6, the reactor was steady at about 8 million watts (thermal power) when the reactivity insertion was made. During the 0.1-second-duration reactivity insertion, the reactor power rose to almost 10 *billion* watts (more than a thousand-fold increase) before descending rapidly to about 200 million watts by the time the SCRAM (chain reaction stoppage) was actuated.

If, in Test No. 6, the increase in SEFOR's power level caused by the positive reactivity insertion had not hesitated as predicted by the Doppler effect; if the reactivity insertion had been accidentally maintained longer than the planned 0.1-second duration; and if the SCRAM procedure had accidentally failed—there was the real possibility that the power level might have increased to levels with the potential to rupture the containment. If that were to occur, there was the potential for some considerable fraction of the 900 pounds of plutonium to be released in aerosol form into the atmosphere.

There was no explosion, and the data obtained from the six experiments that week in December 1971 were considered a demonstration that the Doppler effect could provide an important safety margin for fast-breeder reactors using plutonium-oxide fuel. Within a month, the reactor was closed and placed in SAFSTOR condition (partially decommissioned).

In the concluding chapters, this book considers the "What if" questions posed two paragraphs above. We believe the book provides evidence that the Doppler effect demonstrated by the reactor's plutonium-oxide fuel might have been overridden by autocatalytic effects following fuel melting if the accidental events postulated above occurred in combination during the prompt-critical transient tests concluding with Test No. 6 in December 1971.

We provide documented evidence in chapter 6 that SEFOR suffered a partial SCRAM failure during an experiment in 1970. That SCRAM system failure probably did not seriously endanger the reactor, due to the smaller amount of positive reactivity that had been inserted into the reactor core in that experiment.

We believe that the SEFOR test program provides reasons to seriously consider what might have happened if an identical partial SCRAM (as occurred in 1970) had occurred in any of the prompt-critical transient tests that concluded with Test No. 6 in December 1971.

CHAPTER 3

Nuclear Fission Bombs and Reactors

Nuclear Energy

In the molecules of substances the atoms are held together by electric forces, and the potential energy stored by these forces is known as chemical energy . . . it can be helpful to imagine that an atom consists of a very compact kernel, called the nucleus, which is surrounded by shells of electrons. The nucleus contains particles called neutrons, which are not electrified, and others called protons which carry positive electricity. Neutrons and protons are roughly equal in weight and are nearly 2000 times heavier than electrons. The whole atom is tiny; so tiny than an ordinary person can hardly imagine that it has any size at all. A sheet of paper is about a million atoms thick. The nucleus is very much smaller: it would take about 100,000 nuclei to stretch across one atom. . . . The forces between the protons and neutrons inside an atomic nucleus are millions of times stronger than those between the atoms in a molecule. And so the potential energy due to nuclear forces is very much greater than that due to chemical forces. When one pound of carbon burns it produces 14,500 BTU of heat energy, but its nuclear energy is 39 million million BTU. This vast store of energy in a pound of carbon corresponds to about ten thousand million KWH of electrical energy, which is roughly equal to the entire output of a very large power station in a year.

F. J. M. Laver, 1962[1]

Nuclear Bombs

The object of the project is to produce a practical military weapon in the form of a bomb in which the energy is released by a fast neutron chain reaction in one or more of the materials known to show nuclear fission.

Robert Serber, 1943[2]

Thermal (Slowed) Neutron Fission Reactor
Average Neutron Speed ~2,000 meters/second (~4,500 miles/hour)

Fast Neutron Fission Reactor
Average Neutron Speed ~20 million meters/second (~45 million miles/hour)

A Nuclear Fission Primer

The principal purpose of the SEFOR experiments was to provide data that would *demonstrate* a *theory-predicted nuclear reaction effect* that could be important for a fast-neutron reactor's safe control and shutdown—even under accident conditions. SEFOR did demonstrate the inherent capacity of its plutonium-oxide fuel arrangement to decrease, due to the Doppler effect, the severity of a potential runaway nuclear power excursion (explosion). However, it does not appear to have been the end-all answer to the nuclear explosion hazard that was implied in the public announcements of the experiments.

A satisfactory answer to the question of whether an accidental nuclear explosion could occur in a fast-neutron fission reactor that could release enough energy to fail the containment depends on two primary factors: the explosion magnitude that is possible, and the explosion magnitude required to fail the containment. To consider the question of the explosion magnitude that is possible, either at an experimental reactor similar to SEFOR or in a larger commercial fast reactor that might be proposed for electric utility service, we must provide the reader with an understanding of certain nuclear reaction facts.

Atoms

The idea that all forms of matter are different arrangements of point-like particles separated by space was proposed centuries ago by the Greeks, but evidence for such an understanding of the *nature of things* only gained scientific acceptance about a hundred years ago. Most important to the arguments in this book, we now know that all mass, whether gas, liquid, or solid form, is composed of such "particles." All such particles, which the Greeks called "atoms" (meaning *indivisible*), are extremely small, very sparsely separated in space, and *in constant motion*.

Einstein provided the scientific argument for the existence of atoms in 1905 by explaining the rapid motion of plant-pollen particles suspended in water and observed with a microscope in 1827 by Robert Brown. Einstein showed that the motion of the suspended (pollen) particles, called Brownian motion, was the result of extremely large numbers of collisions of (atomic) "particles" of liquid water with the much larger suspended pollen particles. We now know that the "particles" of water were actually *molecules*—each consisting of two hydrogen atoms combined with one oxygen atom.

Structure of Atoms

By about 1900 (consider how recently!), we had learned that all matter is composed of approximately 100 different atoms, called the "elements," displayed to all chemistry students in the periodic table. And we soon learned that all of the elements are different combinations of just three "elementary" particles:

- Electrons—negatively charged particles with a mass of 9.11×10^{-31} kg, discovered in 1897
- Protons—positively charged particles with a mass of 1.675×10^{-27} kg, discovered in 1920
- Neutrons—neutrally charged particles with a mass of 1.673×10^{-27} kg, discovered in 1932

Atoms are differentiated by the number of protons they contain. The number of electrons is normally equal to the number of protons, so that the atom is neutrally charged; the balance of the atom's mass is comprised of neutrons. This mental picture of the atom is referred to as the planetary model; although now considered to be unrealistic, it will aid our "visualization." We know that the size of the nucleus, which is a very densely compacted collection of protons and neutrons at the center, is an extremely small fraction of the size of the atom (picture a "cloud" of electrons with an outer boundary defining the size of the atom). The diameter of the nucleus (of any atom) is about 100,000 times smaller than the atom's diameter. Consequently, any atom is mostly space. We turn our attention now to *nuclear explosions*.

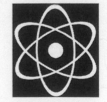

Explosion in (or of) a material is defined as the disassembly (flying apart) of the material that results from very rapid expansion due to increase in temperature (heating) and pressure. *Chemical reaction* explosions are the result of rearrangement of the electrons of the atoms. The nuclei of atoms are not affected in such reactions. This picture is simplified, but it aids us in focusing on reactions that result in changes in the *nucleus*, specifically the nuclear *fission chain reaction*. The discovery of this type of reaction started the world down the path of investigating the potential for making "super" bombs with thousands of times the damage potential of bombs based on chemical reactions. A parallel path of investigation soon appeared in which the goal was to control the rate of nuclear fission reactions to allow the potential release rates of energy to be harnessed for "peaceful" purposes, such as to generate electricity. So began the eventual path to SEFOR.

Nuclear Fission

The nuclear fission reaction was discovered in the laboratory in late 1938 in Germany by Hahn and Strassmann. They were studying uranium to understand what happened if neutrons collided with uranium nuclei. The experimental results indicated that some of the nuclei of uranium atoms were somehow splitting (later called *fissioning*) and forming two new elements that Hahn and Strassmann identified as barium and krypton. This contradicted their expectation that the striking of the uranium atoms by neutrons would either result in the absorption of the neutron or "chip off" small parts of the atoms, changing the makeup of the uranium only slightly. Instead, they were observing the uranium atoms to be splitting into roughly equal parts! Before the year 1939 ended, scientists around the world agreed that the reaction Hahn and Strassmann had observed was explained by the following formula:

1 uranium atom + 1 neutron = 1 barium atom + 1 krypton atom
+ 2 neutrons

It was soon learned that the atoms formed in a uranium fission event would not always be the same; many other pairs of elements than barium and krypton can result. In fact, a large number of *fission products* are produced, many of which are dangerously radioactive. We know that production by a bomb of such radioactive products results in a severe hazard to life that unavoidably accompanies the destructive blast and heat effects of the bomb. We know as well that nuclear reactors produce copious amounts of highly radioactive products as the fuel fissions. The primary hazard considered in this book is the potential for a relatively minor (compared to a bomb) nuclear explosion in a fast reactor that could completely fail the containment and release large amounts of radioactive aerosolized fission products and fuel components (principally uranium and/or plutonium) into the environment.

Scientists soon realized that the combined masses of the fission-product atoms formed is slightly less than the mass of the "parent" uranium atom, and the "missing" mass is converted to energy via Einstein's famous equation, $E = mc^2$. *This knowledge continues to haunt the world.* The nuclear fission reaction results in the conversion of this "lost mass" to kinetic energy (speed) of the products of the reaction—the fission products and the neutrons. The realization quickly followed that if a uranium nucleus were fissioned by a single neutron, and the fission reaction produced two (or more) new neutrons, a "chain reaction" could occur with an *exponentially increasing rate* (doubling the reaction rate at each step). The time step for the doubling would be related directly to the speed of the neutrons producing the chain reaction—the faster the neutrons, the shorter the doubling time (the faster the reaction rate), and the more powerful the explosion. SEFOR was a fast-neutron reactor.

The idea of a fast-neutron fission chain reaction is not complicated. Assume that each fission generates exactly two neutrons, each of which can cause a subsequent fission that generates two more neutrons. The number of neutrons produced at each chain reaction step then increases as two raised to the step number N, or 2^N. Scientists quickly realized that a chain reaction completing eighty such nuclear reaction steps could occur in less than 1 microsecond (1 millionth of a second) and could liberate (explosion) energy equivalent to approximately 20,000 *tons* of TNT (trinitrotoluene, a chemical explosive). The number of neutrons formed in eighty such steps can be calculated with your iPhone: $2^1 = 2$ $2^2 = 4$ $2^3 = 8$ $2^4 = 16$ ... $2^{80} = 1,208,925,819,614,600,000,000,000$.

Eighty such steps of nuclear chain reaction would require fissioning of only about 1 kilogram (2.2 pounds) of uranium, or a sphere with diameter about 3 inches (baseball size). The largest bomb in the arsenals of any nation at that time, their size limited usually by the carrying capacity of the largest military aircraft available, had an explosive yield equivalent to about 6.5 tons (more than 3,000 times smaller) of TNT. *The race was on to build an atomic bomb.*

It has been almost three-quarters of a century since that race began. In many ways, the race to develop controlled nuclear reactors for "peaceful" purposes began at the same time. But those seventy-five years seem in the dim past to the students who are studying at the University of Arkansas today; most of them were born fifty years after the discovery of fission and twenty years after SEFOR had closed. Although most of the students today have "heard of" nuclear reactions, they appear for the most part ignorant of the scientific knowledge required to decide whom to believe on questions as basic as how to maintain the nuclear weapons arsenal in "safe" condition or how to "safely" utilize nuclear power reactors to generate electricity. Both of these questions are at the top of the list in importance to our society today. To begin meaningful consideration of the relevance and importance of the results of the SEFOR experiments, we require an understanding of the basic science differentiating nuclear fission reaction rates that occur in bombs and in nuclear power reactors.

Nuclear Fission Bombs

To provide a brief introduction to the physics of nuclear fission bombs we defer to Robert Serber's *The Los Alamos Primer*.[3] The primer contains the lecture notes prepared for presentation to the Manhattan Project scientific staff assembled at Los Alamos with the mission to build a nuclear fission bomb that could be delivered to its target by military aircraft. The lecture notes, later published as *The Los Alamos Primer,* were written in April 1943 and classified top secret until released to the public in 1965.

In twenty-one succinct parenthetical statements, the primer spells out what we need to know about fission bombs to begin our consideration of the relevance of the SEFOR experiments to the question of the nuclear explosion potential attending operation of fast-neutron reactors. The following numbered statements are direct (partial) quotes excerpted from the primer. As some of the statements of the primer are not required for our purposes, and as our goal here is to reduce the information presented to the minimum required to quantify the potential for explosive energy release in fast reactors, we have emphasized selected statements, some of which are abbreviated. The quotations presented below (with brief explanatory remarks following) are the most important for our purposes. The titles of the remaining statements are included so that the heading numbers correspond directly to the twenty-one numbered statements in the primer.

1. The Object

The object of the project is to produce a practical military weapon in the form of a bomb in which the energy is released by a fast neutron chain reaction in one or more of the materials known to show nuclear fission.

The focus is on the fissile (capable of being fissioned) materials U_{235} (uranium isotope 235) and Pu_{239} (plutonium isotope 239). The scientists at Los Alamos in the spring of 1943 anticipated that realization of a practical bomb would require fast-neutron reactions. SEFOR was a fast-neutron reactor fueled with plutonium.

2. Energy of Fission Process

The direct energy release in the fission process is of the order of 170 million electron-volts (MEV) per atom. This is considerably more than 10^7 times the heat of reaction per atom in ordinary combustion (chemical burning) processes. . . . 1 kg of 25 (fished) is equivalent to about 20,000 tons of TNT.

The physicists at Los Alamos had code expressions for key words and phrases: "fissioned" became "fished," 25 stands for U_{235}, 28 for U_{238}, and 49 for Pu_{239}. 10^7 is ten million. The final sentence changed forever the prospects of military conflict; nuclear fission weapons would be thousands of times more powerful on a weight-for-weight basis than chemical-based bombs.

Release of this energy in a large-scale way is a possibility because of the fact that in each fission process, which requires a neutron to produce it, two neutrons are released. Consider a very great mass of active material, so great that no neutrons are lost through the surface, and assume the material so pure that no neutrons are lost in other ways than by fission. One neutron released in the mass would become 2 after the first fission, each of these would produce 2 after they each had produced fission so in the nth generation of neutrons there would be 2^n neutrons available. Since in 1 kg of 25 there are $5*10^{25}$ nuclei, it would require about n = 80 generations to fish the whole kilogram.

The original primer notes gave the number of nuclei in 1 kg of U_{235} as $5*10^{25}$. Serber changed this number when the primer was published to the correct value—$2.58*10^{24}$. $2.58*10^{24}$ (2,580,000,000,000,000,000, 000,000) is 2.58 million million million million, a number so large as to be difficult to provide meaningful illustration. This point is key to our discussion; the rapid fissioning of 1 kg of U_{235}, which is $2.58*10^{24}$ atoms, would cause an explosion with an energy release equivalent to about 20,000 tons of TNT. It follows directly that if the fissioning process is limited to smaller fractions of 1 kg, the energy release would be proportionally smaller:

Fraction of 1 kg fissioned, %	Energy release, tons TNT
100	20,000
10	2,000
1	200
.1	20
.01	2
.0025 (1/4 of 1/100 of 1%)	0.5 (1,000 pounds TNT)

The estimated maximum nuclear explosion that could be economically/practically contained for a commercial-size (electric) fast-breeder reactor at the time the Clinch River Demonstration Plant was being considered was equivalent to about 1,000 pounds of TNT. It follows that a fast-neutron fission chain reaction of U_{235} of ¼ of 1/100 of 1% of 1 kilogram of U_{235} (*0.025 grams*) would release approximately the same amount of energy as the maximum explosion energy yield that could be realistically (economically) contained.

While this is going on the energy release is making the material very hot, developing great pressure and hence tending to cause an explosion. In an actual finite setup, some neutrons are lost by diffusion out through the surface. There will be therefore a certain size of say a sphere for which the surface losses of neutrons are just sufficient to stop the chain reaction. This radius depends on the density. As the reaction proceeds the material tends to expand, increasing the required size faster than the actual size increases. *The whole question of whether an effective explosion is made depends on whether the reaction is stopped by this tendency before an appreciable fraction of the active material has fished.* (emphasis added)

We have previously defined explosion as the disassembly of the material that results from very rapid expansion due to increase in temperature and pressure. The rapid release of energy (as heat) of the fission chain reaction causes the temperature and pressure to rise rapidly (measured in millionths of a second). The material expands correspondingly rapidly, but precise prediction of the rate of expansion is difficult and uncertain. As the material expands, the density decreases (more space between atoms), and the number of neutrons escaping through the expanding surface increases. If the loss of neutrons by escape through the surface grows too large, the nuclear fission reaction stops. So, an effective explosion requires (1) assembly of a fissile material in an amount sufficient to sustain a chain reaction and (2) holding it together for a period sufficient to cause the desired number of fission chain reaction steps to occur (here illustrated as eighty) before the expansion shuts down the process.

Note that the energy released per fission is large compared to the total binding energy of the electrons in any atom. In consequence, even if but ½% of the available energy is released . . . the temperature is raised to the order of $40*10^6$ degrees. . . . Expansion of a few centimeters will stop the reaction, so the whole reaction must occur in about $5*10^{-8}$ sec, otherwise the material will have blown out enough to stop it. *Now the speed of a 1 MEV neutron is about $1.4*10^9$ cm/sec and the mean free path between fissions is about 13 cm so the mean time between fissions is about 10^{-8}sec.* (emphasis added)

If only one-half of 1% of the energy is released, the temperature of the material rises to about 40 million degrees (centigrade), a temperature more than sufficient to turn any material known to gas. The speed of a 1 MEV neutron is about 1.4 billion cm/s (14 million m/s) or about 4.7% of the speed of light. These neutron speeds, coupled with the distance traveled between fissions, indicates a measure of the average time between fissions. It is about 1/100th of a microsecond, or 1/100 millionth of a second. Neutron speeds of this magnitude are also possible in fast reactors, whereas neutron speeds are about a thousand times slower in thermal (moderated) reactors. This is another key point.

Since only the last few generations will release enough energy to produce much expansion, it is just possible for the reactions to occur to an interesting extent before it is stopped by the spreading of the active material.

Crucial points are coming fast now. The material must be held together (continually squeezed) for a sufficiently long period (extremely small fraction of a second) for the chain reaction to release sufficient energy to cause the desired damaging effect of the bomb. Serber's phrase "interesting extent" refers to the expected damage potential (20,000 tons TNT equivalent). Holding the material together long enough to get the desired energy release was a major challenge facing the bomb designers. This point is also relevant to fast-reactor explosion hazards, because even though the balance between the rate of energy liberation (by fission) and the "shutting down" of that energy release by expansion is expected to result in much less explosion damage than could occur in a nuclear weapon, *the question becomes one of whether the amount of energy yield that would be possible during credible accident conditions could exceed the explosion confinement capability.* As we mentioned, we will see that economic limits are reached for confinement of about 1,000 pounds TNT equivalent yield. *Thus, even a very inefficient explosion yield (from as little as 1/400 of 1% of 1 kilogram, or 0.025 gram) of fissile material could exceed the economic limit for containment construction.*

3. Fission Cross Section

The materials in question are U_{235}, U_{238}, and Pu_{239} and some others of lesser interest. Ordinary uranium as it occurs in nature contains about 1/140 of 25, the rest being 28 except for a very small amount of 24. . . . We see that 25 has a cross-section of about $2.5*10^{-24}$ cm^2 for neutron energies exceeding 0.5 MEV and rises to much higher values at low neutron energies. For 28, however, a threshold energy of 1 MEV occurs below which the cross-section is effectively zero. Above the threshold the cross section of 28 is fairly constant and equal to $0.7*10^{-24}$cm^2.

Serber is focusing on the uranium isotopes—we will return to plutonium later. The *cross sections* can be thought of as measures of the *size* of the nucleus of a fissile atom, and the size as a measure of the probability that a collision will occur between that nucleus and a neutron passing through the atom (mostly space)—the "bigger" the nucleus of the fissile atom, the more probable that a collision resulting in fission will occur. It was learned soon after fission was discovered that the cross section of U_{235} (a measure of the probability of collision with a neutron) is approximately $2.5*10^{-24}$ cm^2 for neutrons with energies above 0.5 MEV (speed approximately 7 million m/s), but it increases greatly at slower neutron speeds. However, the cross section of U_{238}, which comprises about 99.3% of natural uranium, is about $0.7*10^{-24}$ cm^2 for high neutron speeds and is effectively zero for low neutron speeds. The result is that significant amounts of energy release by chain reaction of U_{235} in (natural) uranium fuel (0.7% U_{235} and 99.3% U_{238}) can only occur if the neutrons produced by fission are slowed to speeds of about 2000–4000 m/s (so-called thermal neutrons). At such neutron speeds, very little fissioning of U_{238} occurs. This is a critical point in our discussion. *This fact provides the basis for the oft-repeated statement that "nuclear reactors cannot explode like a nuclear bomb."* Reactors fueled with natural uranium cannot be made to provide fission reaction rates suitable for electrical power generation unless the neutrons produced by fission (and continuing the chain reaction) are slowed to the so-called thermal level. *The italicized statement does not apply to reactors that have significantly enriched fissile content, like SEFOR.*

4. Neutron Spectrum

In Fig. 2 (not included) is shown the energy distribution of the neutrons released in the fission process. The mean energy is about 2 MEV, but an appreciable fraction of the neutrons released have less than 1 MEV energy and so are unable to produce fission in 28.

The important point here is that neutrons with energy less than 1 MEV produce very few fissions of U_{238}, but slow neutrons, on the other hand, readily fission U_{235} (and very importantly, we will see that slow neutrons even more readily fission Pu_{239}).

5. Neutron Number

The average number of neutrons produced per fission is denoted by ν. It is not known whether ν has the same value for fission processes in different materials, induced by fast or slow neutrons or occurring spontaneously. The best value at present is $\nu = 2.2+/-0.2$ although a value $\nu = 3$ has been reported for spontaneous fission.

More is now known about the neutron number for uranium and plutonium. It is greater than 2 for both, and it is greater for plutonium than uranium. Our use of a value of 2 for purposes of estimating fission reaction rates is too low, but will serve our purposes of illustration here.

6. Neutron Capture

When neutrons are in uranium they are also caused to disappear by another process represented by the equation 28 + n = 29 + X. The resulting element 29 undergoes two successive beta transformations into elements 39 and 49. The occurrence of this process in 28 acts to consume neutrons and works against the possibility of a fast neutron fission chain reaction in material containing 28. It is this series of reactions, occurring in a slow neutron fission pile, which is the basis of a project for large scale production of element 49.

Another critical point. Plutonium (element 49) is formed in nuclear reactors by this reaction. Plutonium can be made most expeditiously in fast-neutron reactors—under some circumstances sufficiently fast to produce more fissionable (plutonium) fuel than was used from the beginning of the reaction. Such a process, called "breeding," became a primary goal of the reactor designer, since it would allow the uranium 238 isotope to be converted to plutonium, thus offering the possibility of extending available energy of uranium found in the earth by perhaps a factor of a thousand, stretching the available fissionable fuel supplies from a few decades to several centuries.

7. Why Ordinary U Is Safe

Ordinary U (meaning as-mined), containing only 1/140 of 25, is safe against a fast neutron chain because, (a) only 3/4 of the neutrons from a fission have energies above the threshold of 28, (b) only ¼ of the neutrons escape being slowed below 1 MEV, the 28 threshold before they make a fission. So the effective neutron multiplication number in 28 is $v = \frac{3}{4} \times \frac{1}{4} \times 2.2 = 0.4$. Evidently a value greater than 1 is needed for a chain reaction. Hence a contribution of at least 0.6 is needed from the fissionability of the 25 constituent. One can estimate that the fraction of 25 must be increased at least 10-fold to make an explosive reaction possible.

This paragraph further explains the "impossibility" of a reactor fueled with natural uranium (1/140 isotope 235, the balance 238) exploding like a bomb. The last sentence is particularly important to us; it states that the fraction of the 235 isotope in (uranium) fuel must be increased at least tenfold to make a violent explosive reaction possible. We will see that fast reactors fueled with uranium or plutonium are typically enriched to about 20% fissile material, whereas a tenfold enrichment of natural uranium would give a fissile concentration of about only about 7%. *The result is that the fuel enrichment required in fast reactors decidedly changes the safety picture; fuel enriched to the 20% level definitely has the potential for a violent nuclear explosive reaction.*

8. Material 49

As mentioned above, this material is prepared from the neutron capture reaction in 28. So far only microgram quantities have been produced, so bulk physical properties of this element are not known. Also its ν value has not been measured. Its cross section has been measured and found to be about twice that of 25 over the whole energy range. It is strongly alpha-radioactive with a half-life of about 20,000 years. Since there is every reason to expect its ν to be close to that for U and since it is fissionable with slow neutrons, it is expected to be suitable for our problem and another project is going forward with plans to produce it for us in kilogram quantities. Further study of all its properties has an important place in our program as rapidly as suitable quantities become available.

We can update this statement for the plutonium isotope 239 (Material 49). In addition to the knowledge that the cross section (reflecting the probability of fission of a given plutonium nucleus by a neutron) is about twice as great as U_{235} over the entire neutron energy (speed) range, the currently accepted value of ν is closer to 3, indicating a faster exponential chain reaction rate. We will return later to the subject of radioactivity when we attempt to understand the potential consequences of fast-neutron reactor accidents that could breach the containment and release the highly radioactive fission products produced by operation of the reactor as well as the highly radioactive plutonium remaining in the reactor to the environment. We note here that plutonium is widely considered one of the most toxic materials known. Serber stated that the half-life of Pu_{239} is about 20,000 years. Today the accepted value is known to be closer to 25,000 years.

9. Simplest Estimate of Minimum Size of Bomb

We will skip most of the technical details of this section, as it is not necessary for us to follow the important arguments presented here. We focus on the determination of the minimum size that would allow neutrons to "leak" through the surface sufficiently rapidly that the initial concentration of neutrons would die out rather than build up (stopping the chain reaction). Serber gives two estimates of the minimum size, assuming the bomb material to be pure U_{235} in the form of a sphere. This size (measured as its diameter) is called the *critical* size, and its amount of material (mass) is the *critical* mass. The two estimates given were 9 and 13.5 centimeters, with corresponding masses of about 60 kilograms and 200 kilograms. Today, unclassified data on the critical mass of a U_{235} sphere suggests values slightly smaller than Serber's estimate of 60 kilograms. These values assume no reflection of neutrons back into the fissionable material after they exit the surface. It would be expected that if the fissile material were surrounded by a material that would reflect neutrons escaping the surface back into the material, the critical mass (and radius) would be reduced.

10. Effect of Tamper

If we surround the core of active material by a shell of inactive material, the shell will reflect some neutrons which would otherwise escape. Therefore a smaller quantity of active material will be enough to give rise to an explosion. The surrounding case is called a tamper.

More current values for critical mass (in kg), with spherical geometry, for weapons-grade fissile materials are:

	U_{235}	Pu_{239}
Bare	56	11
Thick Uranium Tamper	15	5

The bottom line is that a "tampered" sphere of essentially pure Pu_{239} weighing about 5 kg (about 11 pounds), of which about 20% (~1 kg) fissioned, delivered about 20,000 tons TNT equivalent explosive power at Nagasaki in 1945. The bomb's plutonium sphere was about the size of a baseball. The amount that fissioned was approximately the size of a golf ball, and the amount of mass that was converted to energy by Einstein's equation, $E = mc^2$, was approximately 1 gram (1 pound mass is 454 grams).

11. Damage

Several kinds of damage will be caused by the bomb. A very large number of neutrons is released in the explosion. One can estimate a radius of about 1,000 yards around the site of explosion as the size of the region in which the neutron concentration is great enough to produce severe pathological effects. Enough radioactive material is produced that the total activity will be of the order of 10^6 curies even after 10 days. Just what effect this will have in rendering the locality uninhabitable depends greatly on very uncertain factors about the way in which this dispersion by the explosion occurs. However, the total amount of radioactivity produced, as well as the total number of neutrons, is evidently proportional just to the number of fission processes, or to the total energy release. The mechanical explosion damage is caused by the blast or shock wave. . . . If destructive action may be regarded as measured by the maximum pressure amplitude, it follows that the radius of destructive action produced by an explosion varies as the cube root of the Energy (yield). Now in a ½ ton bomb, containing ½ ton of TNT, the destructive radius is of the order of 150 feet. Hence in a bomb equivalent to 100 kilotons of TNT, one would expect a destructive radius of the order of . . . about 2 miles.

We know now that the damage produced by a nuclear fission bomb is of three kinds: blast, thermal (heat), and radiation. The plutonium "test" device exploded at the Trinity site (essentially identical to the bomb dropped on Nagasaki) was mounted on a steel tower and exploded at a height of about 100 feet above ground level. The division of the total yield for such kiloton-range weapons among the three categories is, approximately, blast 50%, thermal (heat) 35%, and radiation 15%. While nuclear fission weapons were principally designed to deliver damaging blast effects, the thermal energy released is of such magnitude as to cause extreme temperatures—sufficient to vaporize most materials near the explosion (the steel tower at the Trinity site was vaporized) and cause severe burns to unprotected persons at considerable distances as well as starting fires that under certain conditions can reach firestorm proportions. The radiation hazards of nuclear weapons, both primary effects of the bomb and secondary effects of the large amounts of radioactive materials released into the environment, have assumed much more importance as we have learned more about the overall effects of nuclear weapons.

12. Efficiency

As remarked in Sec. 3, the material tends to blow apart as the reaction proceeds, and this tends to stop the reaction. In general, then, the reaction will not go to completion in an actual gadget. The fraction of energy released relative to that which would be released if all active material were transformed is called the efficiency.

"Gadget" was the scientists' code for bomb. "Transformed" here means fissioned. We will see that this definition of "efficiency" of nuclear explosive energy release can be usefully extended to the fissile contents of a fast (enriched fuel) nuclear reactor. *Consideration, then, of the possibility that some fraction of the active (fissile) material in the reactor can be fissioned under conceivable circumstances, including accidents and natural disasters, with sufficient energy release to fail the reactor containment becomes the question that brought us here.*

13. Effect of Tamper on Efficiency

For a given mass of active material, tamper always increases efficiency. It acts both to reflect neutrons back into the active material and by its inertia to slow the expansion, thus giving opportunity for the reaction to proceed farther before it is stopped by the expansion.

The quantitative determination of tamper effect is complicated, but all we need to take away here is that if the fuel in a fast reactor were to suffer a reactivity excursion, the reactivity might be further increased in severity by the tamper effect of the masses of material surrounding the fuel. We can see that this is an important, but uncertain, part of the puzzle of quantifying the expected explosion severity in a fast-reactor accident.

14. Detonation

Before firing, the active material must be disposed in such a way that the effective neutron number is less than unity. The act of firing consists in producing a rearrangement such that after the rearrangement the neutron number is greater than unity. This problem is complicated by the fact that, as we have seen, we need to deal with a total mass of active material considerably greater than the critical in order to get appreciable efficiency.

This paragraph restates the basic requirement for achieving, and maintaining for a sufficient time, an exponentially increasing fission chain reaction. First a mass must be assembled that is configured (arranged) in such a way that is subcritical. The bomb dropped on Hiroshima accomplished this by assembling the (potentially critical) U_{235} in two (separated) sections. The bomb was detonated by rapidly combining the two masses; this was accomplished via the "shooting" method in which the two masses, separated in a tube (gun barrel), were rapidly combined by explosively driving one of the masses into the other. Simultaneously, the bomb released a strong source of neutrons at the center of the critical mass that initiated the bomb chain reaction. If during the assembly of the critical mass there are reaction-initiating neutrons present, a reaction can begin, termed a fizzle, that will not be exponentially increasing to produce the desired energy yield. This was a potential problem for the early bomb designers when plutonium was used as the fissile material because of the effectively higher reaction rates that controlled the fission rate of the plutonium. The reaction rates were too high to allow combination of the two masses of the fuel with the "shooting method"; the combination of the separate masses by shooting was just too slow. Consequently, the plutonium bomb utilized a sphere of plutonium that was subcritical. The fissile mass was "rearranged" by squeezing it using powerful explosives that spherically crushed the bomb material, increasing its density beyond the point of criticality, at which time neutrons were introduced that initiated the chain reaction fission process.

The fast-reactor safety problem we are considering is quite different. In thermal reactors (without enriched fuel), although there can be present many times the amount of fissile material in the reactor required for criticality, it is separated into noncritical masses in the reactor so that there is no way, as long as the configuration is maintained, that a reaction can be initiated that will have bomb-like efficiency. However, the potential for a much weaker nuclear explosion in a reactor with enriched fuel is very real. The question is whether the strength of the explosion that might be possible under any conceivable circumstances would be sufficient to fail the containment. We are again back to the question that brought us here.

15. Probability of Detonation (not included)

16. Fizzles

The question now arises: what if by bad luck or because the neutron background is very high, the bomb goes off when the neutron number is very close to zero? It is important to know whether the enemy will have an opportunity to inspect the remains and recover the material. We shall see that this is not a worry; in any event the bomb will generate enough energy to completely destroy itself.

A "fizzle" describes the situation where "firing" of the bomb does not result in a sustained exponentially increasing rate of fission chain reaction. The designers worried that the fraction of the fissile material that fissioned could be so small as to obviate its bomb-utility. However, an important point arises here in our consideration of the explosion potential in fast reactors containing enriched fuel. Serber says there is no "worry" that the fizzled weapon could be inspected (giving up the secrets of its design), because even a very small, fractionally efficient bomb would still be powerful enough to completely destroy the weapon. We have already noted that an explosion yield of 1/100 of 1% of a 20-kiloton bomb like that dropped on Nagasaki would be equivalent to about 4,000 pounds of TNT. The published literature shows that the maximum planned explosive containments provided for any of the fast reactors considered by the United States was less than half that amount.

17. Detonating Source (not included)
18. Neutron Background (not included)

19. Shooting (not included)

20. Autocatalytic Methods

The term "autocatalytic method" is being used to describe any arrangement in which the motions of material produced by the reaction will act, at least for a time, to increase the neutron number rather than to decrease it. Evidently, if arrangements having this property can be developed, they would be very valuable, especially if the tendency toward increasing the neutron number was possessed to any marked degree.

Both the Hiroshima and Nagasaki bombs utilized autocatalytic methods. The shooting method assembles the two subcritical masses sufficiently rapidly that the material is held together for a sufficient time for the energy release to meet the bomb design requirements. The "implosion" method used in the Trinity test device and in the Nagasaki bomb squeezed the bomb core to a fraction of its original volume, and held it there for a sufficient time with chemical explosives aided by the tamper effect. The result was similarly successful for both bomb designs; criticality was reached when the density of the material (mass of bomb material divided by its volume) reached a degree of fission efficiency required to provide the desired explosive yield before the neutrons escaped through the bomb core surface sufficiently fast to shut the reaction down. We will see that autocatalytic processes can result in fast reactors in accidents or as the result of natural disasters. We return to the question that brought us here. The subject of autocatalytic processes will be further considered in chapter 5.

21. Conclusion

From the preceding outline we see that the immediate experimental program is largely concerned with measuring the neutron properties of various materials, and with the ordnance problem. It is also necessary to start new studies on techniques for direct experimental determination of critical size and time scale, working with large but sub-critical amounts of active material.

Serber's conclusions are a short list of the needs then considered to be of high priority for producing a fission bomb. Not surprisingly, these needs were similar to the needs of those people who were already working on the methods for designing fission reactors using the same materials, uranium and plutonium, to generate electrical power.

The First Plutonium Atomic Bomb

The first plutonium atomic bomb ever tested, rigged as a device named "The Gadget" and mounted on a steel tower approximately 100 feet above the ground, was demonstrated successfully in the Trinity test in New Mexico, July 16, 1945. The bomb design was based on the mathematical prediction of a nuclear fission chain reaction initiated in the center of a sphere of plutonium. Mathematical (computer) predictions indicated that if the chain reaction proceeded through about eighty fission generations, an explosive-energy release (yield) of about 20 kilotons (TNT equivalent) would be achieved. This prediction of the explosion yield of a spherical-shaped plutonium mass led the designers to shape the fissile material (plutonium) as a sphere with a hollow central core into which the neutrons could be released to initiate the fission process. The sphere of plutonium was sufficiently large that it could be made "supercritical" by compression (decreasing its volume) using a chemical explosive blanket surrounding the sphere to direct explosive power

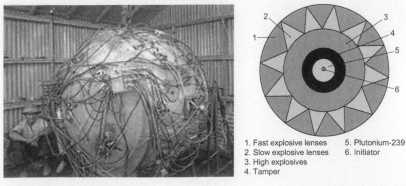

1. Fast explosive lenses   5. Plutonium-239
2. Slow explosive lenses   6. Initiator
3. High explosives
4. Tamper

Left: Trinity Test Device, "The Gadget." *Right:* Schematic of the Gadget.

uniformly on the surface of the sphere and directed (focused) at the center. A "blanket" of explosives symmetrically surrounding the fissile material (the "bomb core") was designed to squeeze the sphere to a sufficiently smaller volume that would make the resulting mass super-critical. Then, at exactly the right instant, a cache of neutrons would be released at the center of the sphere. All of this "supercritical assembly" of the fissile material (plutonium sphere) had to be accomplished in a total time, starting with the detonation of the explosive blanket, of the order of 1 millionth of a second. Perhaps most importantly, the "rearrangement" of the plutonium molecules was designed to cause a perfectly symmetrical decrease in the diameter of the spherical shape during the "squeezing" process. This required an extraordinary capability for controlling the timing and direction of the chemical-explosive blanket detonation wave directed symmetrically inward toward the core center. In the end (mid-year, 1945), after a sufficient amount of plutonium was finally available, requiring nearly two years in the most expensive scientific/industrial undertaking ever attempted at the time, the design and demonstration of the bomb hinged on the ability of the designers to effect the squeezing of a plutonium sphere, maintaining its spherical shape, to a supercritical volume in around a millionth of a second and then release at that instant the neutrons at the bomb core center. If that could be accomplished, the bomb would do the rest. We know the result; within a few days, a device (dressed as a bomb) that is thought (the actual design is still classified) to be essentially identical to the Trinity device, was detonated about 1,600 feet over the aiming point in Nagasaki, Japan.

The burdens of uncertainty undertaken by the bomb designers were, although daunting, limited to barely manageable proportions by designing a process in which a carefully assembled amount of plu-tonium could be, under controlled conditions, made "critical" at an appointed instant when the nuclear reaction would be triggered by the release of fast neutrons at the center of the core and the nuclear explosion would take its course. The predictions required by the sci-entists to design a bomb in which such an exacting process would occur on demand, lasting on the order of a millionth of a second, required an extraordinarily expensive effort by the best scientists in the world.

Relevance of the "Gadget" Design to Explosion Potential in Fast Reactors

The process of constructing the first nuclear (plutonium) fission bomb with nominal 20-kiloton TNT equivalent explosion damage potential, anticipated in Serber's primer, follows relatively simple directions:

- Obtain a sufficient quantity of pure fissile material, approximately 5 kilograms of Pu_{239}. In order for this quantity to be safely handled, it must be separated into at least two parts with distance between the parts sufficient to prevent the volume of the material from becoming arranged compactly (by any means, including accidents) enough to reach super-criticality—the condition where the rate at which new neutrons are being produced in the material just exceeds the rate at which neutrons disappear by absorption (by other bomb materials) or by leakage through the fissile material's surface.
- To effect an explosion, the separated parts (the molecules less densely spaced than required for criticality) of the plutonium must be assembled so as to very rapidly increase the overall density sufficiently to make the material exceed the criticality condition. This is normally accomplished in a plutonium bomb by (extremely rapidly) compressing a single subcritical mass to decrease the volume sufficiently for the mass to become supercritical.
- Just after the instant of assembly, release a collection of fast neutrons at the center of the critical mass, while preventing (to the extent practicable) the expansion of the volume of the material by the extreme heating process that ensues. It is not possible to completely prevent such expansion, but the expansion can be delayed to cause the number of fissions required to produce the design yield of the bomb before the expansion process stops the fission-energy release.

Extreme precision is required for the arrangement of the various parts of the device. In an implosion bomb, which we focus on, the compression and explosion development must be spherically symmetric. To produce such a precisely controlled nuclear reaction process required precision of construction and design that was unheard of in the mid-1940s. Extreme precision is also required in the timing of the bomb detonation processes. The duration of these processes is of the order of a small fraction of a microsecond (1 millionth of a second). In a properly operating bomb, the processes are so rapid that there is virtually no possibility of controlling the process in order to slow or stop it (save the certain expansion of the fuel and fission products as a result of extremely high temperatures).

A nuclear fission bomb is designed to effect a nuclear chain reaction in a collection of fissile material that grows exponentially (in time), releasing a very large amount of energy in an extremely short time period (fraction of a microsecond). The fissile material is "assembled" as a "supercritical" mass, and then a cache of neutrons is released in the interior of the mass. The mass must be "held together" long enough (fraction of a microsecond) for the fast chain reaction process to cause the desired fission energy release (say eighty fission generations that would release approximately 20,000 tons TNT equivalent energy). The desired result is the same as for any bomb—the release of such a large amount of energy in a small volume generates extreme temperatures (measured in millions of degrees in a *nuclear* bomb), and such temperatures transform the materials of the bomb to gas that expands, producing extreme pressures causing "disassembly" of the atoms of the fuel (as well as surrounding materials). Disassembly is another word for explosion. There are other damaging effects (such as radioactivity), but here we focus on the effects that occur in the surroundings as a result of the absorption (by the bomb materials) of the heat liberated in the chain reaction and the damaging pressures that are produced.

In contrast, a nuclear fission reactor for producing electric power is designed to effect a tightly controlled nuclear chain reaction in a collection of fissile material at a constant rate that can be converted to electric power. A typical 1,000-megawatt (one thousand million watts) power station must generate about 3,000 megawatts of heat at a steady rate to convert (roughly) one-third of that energy (power = energy per unit time) to produce 1,000 megawatts of electricity. The nuclear fission reactor in the plant produces the heat energy at a controlled rate and transfers that heat to the reactor coolant, which in turn transfers the energy (as steam) to an electric power generator. The sole purpose of the reactor is to produce the thermal energy (heat) that is carried away from the reactor (to the steam generator) by the reactor coolant.

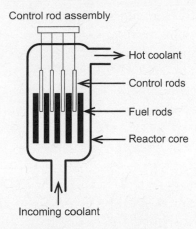

A schematic diagram of a nuclear fission reactor.

In contrast to a bomb, a power reactor must have a coolant fluid that flows continuously through voids (channels) in the fuel. The coolant flow is designed to remove the heat of the fission process at the same (balanced) rate that it is produced. The process of constructing a nuclear fission reactor, then, also appears to follow relatively simple directions. As in a bomb, there is a requirement for extremely precise physical arrangement of the various parts of the device. In a reactor, the arrangement of the fuel must be compartmented to allow passage (through the fuel assembly) of the material (coolant) that *absorbs the heat at the same rate that the reactor core produces it* by nuclear reaction. The nuclear reaction process is controlled with special materials inserted into (or surrounding) the compartmented fuel assembly that absorb neutrons produced in the fission process. The position of these materials (the "control rods") is adjusted continuously (normally automatically) in the core of the reactor so as to balance the rate of generation of neutrons with the rate at which newly produced neutrons are used to maintain the desired fission rate as well as the neutrons that are absorbed (but do not cause fission) in the core or are lost by leakage from the core. The process is a highly fine-tuned balancing process that must be maintained very accurately and precisely. If the reactor speeds up unexpectedly, which it can do if the spatial arrangement of the fuel changes only slightly (to a denser configuration), the heat produced by the fission process can overpower the coolant heat removal capacity. If the rate of heat production exceeds the heat absorbed by the coolant by too large an amount, the temperature of the fuel can exceed the melting (and later vaporization) temperature in an extremely short period. Therein lies the potential for an explosion that might have the potential to fail the containment. And therein lies the problem of assuring an "engineered" design of the plant that will prevent a nuclear explosion of sufficient intensity to destroy any last-barrier containment provided. Engineering safety of such devices is extremely difficult if there is the potential for accidental "rearrangements" of the fuel that might result in very short duration productions of energy capable of destroying the reactor containment. It is simply a matter of having insufficient time to react defensively; the safety margin can be very thin. *The primary purpose of this book is to argue that the proposed use of fast-fission reactors for electric power production is a move in a direction that could result in accidental increases in energy release that reduce the safety margin of control to an unacceptable level.*

The development of nuclear fission reactors for "peaceful" purposes, including electrical power generation, proceeded in parallel with the development of more efficient and powerful nuclear fission weapons systems. As fissile uranium (U_{235}) and plutonium (Pu_{239}) were exceedingly scarce when World War II ended in 1945, the production of plutonium in nuclear reactors suggested the so-called "breeder" reactor, which, while producing electrical power at a controlled rate, could produce fissile plutonium (as a by-product) at a rate greater than the rate that the fissile material was used up to produce power. Producing plutonium with reactors was the primary purpose of the first so-called "production" reactors that the AEC operated during the 1950s and 1960s to make plutonium for nuclear weapons. Nuclear power reactors were soon envisioned to produce plutonium at a rate faster than the fissile uranium was fissioned ("burned"), resulting in accumulation of fissile material faster than it was used in the reactors; hence the reactors would "breed" plutonium by converting the non-fissile uranium (U_{238}) to fissile plutonium. This would mean that the very scarce fissile uranium available could be used while vastly increasing the available fission fuel supply—explaining why the first nuclear reactor that generated electrical power, constructed at the National Reactor Testing Station (NRTS) in Idaho in 1951, was designed to demonstrate the practicability of a reactor that would "breed" new fuel faster than it was being used up in the reactor. That first experimental breeder reactor (*EBR-I*) was a fast-neutron reactor fueled with enriched uranium built for that purpose.

To provide a brief introduction to the physics of nuclear fission reactors, with emphasis on fast-fission reactors, we defer to chapter 10 of Dr. Richard Webb's book *The Accident Hazard of Nuclear Power Plants*, published by the University of Massachusetts Press, 1976:

The Explosion Hazard of the Advanced "Breeder" Reactor (LMFBR)

The liquid metal-cooled, fast neutron, breeder reactor (LMFBR) is an entirely different power reactor concept than the water-cooled reactor. It is designed specially to produce or "breed" fissionable material as a by-product, namely, plutonium fuel, by certain nuclear reactions in the reactor that convert plentiful uranium-238, a weakly fissionable species of uranium, into plutonium. The objective of the LMFBR is to produce more plutonium than is consumed in an operating cycle (about 7% more per year). The excess fuel can then be used to start up other LMFBRs and to fuel the PWRs and BWRs when the useful reserves of rare fissionable uranium (U-235) are depleted, which is estimated to occur in about thirty years.[4] Hence the PWRs and BWRs will ultimately depend on the LMFBR. (The reserves and present stockpiles of U-238 would last the U.S. for a thousand years or so, using the combination of LMFBRs and the water reactors.)

The AEC has projected that about one thousand large LMFBRs would eventually be built, along with a like number of water-cooled reactors.[5] A license application to build the "LMFBR Demonstration Plant" in Tennessee is presently pending; and a smaller LMFBR-like reactor, called the Fast Flux Test Facility (FFTF), is under construction, which is to be used to test LMFBR fuels under LMFBR core conditions encountered in normal operation. The FFTF differs from an LMFBR in that the plutonium-fueled core is not surrounded by uranium 238 for breeding purposes; otherwise, it is basically the same as an LMFBR, from the standpoint of accident hazards.

Unfortunately, the LMFBR has a power excursion (nuclear runaway) potential—indeed, a potential for nuclear explosion as distinguished from a steam explosion—which is even more serious than that of the water-cooled reactors.[6] The LMFBR explosion potential has extremely grave implications. Especially because such a nuclear explosion would produce radioactive plutonium dust, which is extremely toxic. A steam explosion, or more accurately, a "coolant vapor explosion," is defined as the explosive vaporization (boiling) of coolant due to the contact of the coolant with extremely hot, molten fuel. This appears to be the only possible mode of explosion in the water-cooled reactors. A nuclear explosion, on the other hand, is defined as an explosive vaporization of the fuel itself, which involves higher temperatures and potentially a much stronger explosion than a coolant vapor explosion. An LMFBR could also produce strong coolant vapor explosions upon melting, which could add to the explosion force or by themselves be dangerous. As will be discussed, nuclear explosions of the order of 20,000 pounds TNT-equivalent are theoretically possible. For comparison, the maximum economical containment capability is about 1000 pounds TNT.[7] Thus, such an explosive power excursion would vaporize the entire core, rupture the containment, and blow, say, half of the core, amounting to tons of radioactive plutonium, into the atmosphere and boil off practically all of the fission products and blow them into the atmosphere as well. Moreover, the core vaporization process presumably would dispense the plutonium and fission products in the form of a superfine dust, causing severe, geographically widespread, ground contamination.

The consequences of such a heavy product release were estimated in chapter 1 (of Webb). The additional consequences due to plutonium release could be even worse. Because of the extreme toxicity of the plutonium, its release (assuming two tons) and its long life (up to 600 to 24,000 years "half-life," depending on the isotopic species of plutonium) could cause permanent abandonment of over 150,000 square miles (an area the size of Illinois, Indiana, Ohio, and half of Pennsylvania combined). This estimate is based on simply substituting plutonium for the fission products in the atmospheric dispersal-fallout calculation of the WASH-740 report (that is, no extrapolations)[8] and finding that the ground contamination level at the boundary of the 130,000 square-mile zone exceeds a proposed contamination limit of one microgram of plutonium 239 isotope per square meter.[9] Actually, an LMFBR core will contain other, more radioactive, isotopes of plutonium, such that there will be about three to eight times more Pu 239—equivalent radioactivity than if the core were 100% Pu 239.[10] Hence, the equivalent Pu 239 contamination of the 150,000-square-mile zone would be three to eight micrograms/m^2 level proposed as a contamination limit by Willrich and Taylor, who stated that any ground contaminated above one microgram/m^2 "would be likely to be deemed unacceptable for public health,"[11] and also exceeds the Rasmussen Report's contamination limit of three micrograms/m^2, above which "relocation" is to be required.[12]

Incidentally, the above use of the WASH-740 calculation does not involve those aspects of the WASH-740 analysis which differ with the Rasmussen Report; the differences appear to arise in the assumptions of the radioactive fallout dust particle size and contamination limits of Sr 90 following a meltdown accident, and not in the atmospheric dispersal-fallout aspects, were the dust particle size the same between the two reports. This assumes that a severe nuclear explosion of an LMFBR core would generate dust particles of one micron size or less, which is the basis for the 150,000 square-mile ground contamination value in WASH-740. In view of the extreme temperature of such a nuclear explosion, the assumption seems appropriate.

If the "hot particle" theory for lung cancer induction by plutonium dust, as proposed by Tamplin and Cochran and by Geesaman,[13] is correct, then the cancer probability would be 100% for anyone attempting to live in the 150,000 square mile (or greater) fallout zone of the conceived LMFBR explosion. Furthermore, the plutonium dust, because of its extremely long half-life, would forever be in the environment. Presumably, it would be mixed with ordinary dust, kicked up by wind erosion and farming, and blown about and spread by the winds to present a continuous and permanent lung-cancer and other health hazard for any inhabitants of the contaminated and adjacent land. It is clear, therefore, that the question of the power excursion potential of the LMFBR is extremely serious. We shall now examine the state of the science of predicting the LMFBR explosion potential.

The Basic Theory of Nuclear Explosion in LMFBRs

The LMFBR reactor and coolant system closely resemble the Pressurized Water Reactor (PWR) except that the fuel rod bundles are each contained in a coolant channel or "duct" as in a Boiling Water Reactor (BWR). Also, the batch of fuel rods which comprise the core is surrounded by a thick outer ring of rods containing uranium 238 for breeding plutonium, and of each core rod, only the middle vertical section actually contains the fuel, with the top and bottom containing U 238. Hence, the core is completely surrounded by a "blanket" of U 238 rods. The coolant ducts containing each bundle of rods run the full length of the rods.

The reactor physics and the power excursion theory for an LMFBR are similar to those for a water-cooled reactor; except that liquid metal (heated sodium) is used as a coolant instead of water. The absence of water means that the energetic (fast) neutrons emitted by the atomic fissioning process are not slowed down within the core (it turns out that the fast neutrons enable the breeding process to work); but since fast neutrons are less effective in causing fissioning, the concentration or "enrichment" of fissionable material in the fuel must be higher in an LMFBR to achieve a critical reactor—over five times higher than a PWR or BWR. This higher fuel enrichment means that should the LMFBR fuel rods be compacted, either by a meltdown or by an explosion which compresses part of the core, the reactivity might not decrease, as it would in a water-cooled reactor, but could increase.[14]

The reason for this reverse reactivity effect is that in a water reactor the fuel material, being less enriched in fissionable material, cannot sustain a fission chain reaction (criticality), even if fully compacted, unless the water is present between the fuel rods to slow down the neutrons and thereby increase their effectiveness for causing fission. Fuel compaction in a water-cooled reactor would squeeze out water and thus would reduce the reactivity, that is, shut down the fission chain reaction. On the other hand, the compaction of concentrated LMFBR fuel will increase the reactivity, since the nature of highly enriched fissionable material is such that it can be made critical if enough of it is brought together, which the compaction process accomplishes. (An atomic bomb is detonated by compacting extremely rapidly a mass of highly enriched fissionable material with a surrounding TNT-like charge.)

What makes the compaction problem especially serious is that only about 2% core volume reduction, such as could easily occur upon core melting, would raise the reactivity to above the delayed neutron fraction (which is about 0.3% for plutonium-fueled LMFBRs) and thus trigger a power excursion.[15] Yet the potential for fuel compaction in an LMFBR is large, since only about 50% of the core volume is taken up by fuel rods, and the rest by the coolant. Hence, the core compaction potential is over 50%,[16] should the coolant be expelled or drained, leaving a void for fuel to enter; again, fuel compaction occurs whenever fuel fills voided coolant space between the fuel rods. Also, upon melting, the fuel rods of the core would lose their rigidity, and the fuel would then be easily compressible, as by gravity compaction.

Furthermore, it turns out that the rapid expulsion of the liquid metal coolant, as in boiling, can increase the reactivity in an LMFBR as well, due to complicated nuclear effects. This is in contrast with water-cooled reactors, where a loss of coolant will at least reduce the reactivity and thereby shut down the fissioning. Finally, due to its size, a large LMFBR core will contain several "critical mass" loads of fuel, if fully compacted, so that an explosion due to an initial power excursion might rapidly compact a region of the core enough to make it prompt critical, thereby setting off a secondary, more severe excursion. In general, the more rapid the core compaction, the greater the rate of reactivity rise and the resultant power excursion.

In short, the LMFBR is prone to autocatalytic reactivity accidents—that is, the reactor is its own catalyst (for generating power excursions, since fuel overheating due to some malfunction can cause a core meltdown or cooling boiling, which in turn could conceivably generate disastrous secondary excursions by some rapid recompaction process before the fissioning would be finally stopped by "core disassembly"—blowing the core completely apart by explosion.

Incidentally, the LMFBRs, like the water-cooled reactors, are being designed with a negative Doppler reactivity effect which can safely terminate minor power excursions caused by slight reactivity rises above the prompt critical reactivity level, the threshold for power excursion.

The Doppler effect has been demonstrated for LMFBRs in power excursion experiments using a small LMFBR reactor called SEFOR. However, the AEC's characterization of these SEFOR tests can be very misleading. Said the AEC in their Proposed Final Environmental Statement for the LMFBR program—"In some of the experiments in SEFOR, the reactivity was intentionally increased well beyond prompt critical, and the rapid transient that resulted was controlled by the negative Doppler reactivity effect." [17] This statement can be taken to imply that SEFOR proved that the LMFBR can tolerate strong reactivity rises—"well beyond prompt critical." But this is not true, for the reactivity in the SEFOR tests was barely raised beyond prompt critical. Numerically, it was only .06% (% reactivity units) beyond prompt critical,[18] compared to, say, 1% for a severe power excursion.[19] Moreover, the Doppler strength in SEFOR was made three to four times greater than it would be in an LMFBR accident situation.[20]

With this background, we are ready to more carefully consider the question of the potential for a fast-neutron fission reactor to suffer, as a result of accident or natural-disaster-caused damage, a nuclear explosion of sufficient power to destroy the reactor containment and allow a catastrophic release of radioactive fuel and fission products to the environment. It seems appropriate to first consider carefully the worst-case implications of such a release.

Notes

1. F. J. M. Laver, *Energy*, Oxford University Press, 1962.

2. Robert Serber, Los Alamos Site "Y," April 1943.

3. Robert Serber, *The Los Alamos Primer: The First Lectures on How to Build an Atomic Bomb*, Berkeley University Press, 1992. Original notes available at https://fas.org/sgp/othergov/doe/lanl/docs1/00349710.pdf

4. "Breeder Reactors," in AEC Booklet, *Understanding the Atom* (1971).

5. WASH-1184, "Cost-Benefit Analysis of the U.S. Breeder Reactor Program" (Jan. 1972), USAEC, pp. 34–39, See Hearings, app. 9.

6. ANL-7657, C. Kelber et al. "Safety Problems of Liquid-Metal-Fast Breeder Reactors" (Feb. 1970); R. E. Webb, "Some Autocatalytic Effects during Explosive Power Transients in Liquid Metal Cooled, Fast Breeder, Nuclear Power Reactors (LMFBRs)" (PhD diss., Ohio State University, 1971): obtainable from University

Microfilm, Inc., Ann Arbor, Mich., no. 72–21, 029; see Dissertation Abstracts, 33, no. 2 (Aug. 1972): 754B–755B. See also chap. 1, nn. 1, 2.

7. See "Safety Criteria," no. 7, of K. P. Cohen and G. L. O'Neill, *Safety and Economic Characteristics of a 1,000 MWe Fast Sodium-Cooled Reactor Design,* in ANL-7120, *Proceedings of the Conference on Safety, Fuels, and Core Design in Large Power Reactors* (Oct. 11–14, 1965), pp. 185–86.

8. Substituting Pu for "fission products" in fig. 14 of WASH-740, p. 70, and table 3, p. 100.

9. Willrich and Taylor, *Nuclear Theft,* Ballinger Publishing Company, p. 26.

10. ANL-7520, *Proceedings of the International Conference on Sodium Technology and Large Fast Reactor Design,* Nov. 7–9, 1968, pt. 2, p. 351. See also ORNL-NSIC-74, B. P. Fish et al., *Calculation of Doses Due to Accidentally Released Plutonium from an LMFBR,* Nov. 1972, p. 5.

11. Willrich and Taylor, *Nuclear Theft,* Ballinger Publishing Company, p. 26.

12. Ras. Rpt., app. 6, p. 66.

13. A. R. Tamplin and T. B. Cochran, *Radiation Standards for Hot Particles,* Natural Resource Defense Council, 1710 N St., N.W., Washington, D.C. 20036.

14. Webb, "Some Autocatalytic Effects," p. 1.

15. ANL-7532, G. H. Golden, *Elementary Neutronics Considerations in LMFBR Design* (March 1969), pp. 55–56.

16. Webb, "Some Autocatalytic Effects," p. 1.

17. PFES, 2: 42–113.

18. GEAP-10010-31, AEC R&D report, SEFOR Development Program, 31st and Final Report (Feb. 1972), pp. 2-1, 2-2, 5-1–5-5, 6–68. See also GEAP-13929, *SEFOR Experimental Results and Applications to LMFBRs* (Jan. 1973), pp. 1-1, 2-11, 4-1. See also Hearings, p. 181.

19. HEDL-TME-71-34, D. Simpson et al., *Assessment of Magnitude Uncertainties of Hypothetical Accidents for the FFTF* (Mar. 27, 1971), p. 38.

20. GEAP-10010-31, p. 5-1.

CHAPTER 4

Catastrophic Release of Radioactive Materials

In an accident involving a plutonium reactor, a couple of tons of plutonium can melt.

Edward Teller, 1908–2013

Before considering further the potential for catastrophic releases of radioactive materials from reactors designed for generating electricity, we should quantify the hazards that could have been realized as a result of accident at SEFOR. The magnitude of the radioactivity hazards that attended the operation of the SEFOR experimental reactor should not be considered comparable to commercial nuclear reactors, then or now, with two exceptions.

Most importantly, SEFOR was a fast reactor operating with enriched plutonium fuel. The heightened potential for a large-scale release of aerosolized plutonium to the atmosphere posed a new and highly contentious risk to the public—because of plutonium's potential use as bomb material and its reputation for extreme radiotoxicity if it gains entry to specific organs in the human body. We will return to the plutonium question, but first we should quantify the risks at SEFOR of the fission-product materials that accumulated in the reactor.

The largest quantity of radioactive fission-product material contained in SEFOR at any time during its operation was a small fraction of that contained in typical commercial power reactors such as the two ~1,000-megawatt (electric) plants operating today on the Arkansas River some 80 miles distant. SEFOR's fuel content was much smaller, and SEFOR was operated only intermittently—to meet the specific needs of the research program designed to address questions relating directly

to public safety (primarily the demonstration of the safety feature associated with a negative Doppler coefficient). During most of SEFOR's
approximately three-year operating period, the reactor was charged
with approximately 380 kilograms (836 pounds) of plutonium and (in
total) generated 25,764 megawatt-hours of thermal energy (heat) during a total operating time of 3,895 hours (162.3 days). In contrast, a
single 1,000-megawatt (electric) reactor, "burning" uranium, operating
steadily for three years (typical period between refueling) would produce approximately 3,500 x 3 x 365 x 24 = 92 million, megawatt-hours
of thermal energy (heat). Approximating the thermal energy released
for each plutonium nucleus fissioned to be 200 MeV, the number of
fuel nuclei fissions that would have occurred in SEFOR compared to a
single commercial reactor during a three-year period would be:

SEFOR	3 million billion billion
1,000 MW Plant	10 billion billion billion

The amount of radioactive fission products produced, which is
directly proportional to the number of fissions occurring, is more than
3000 times less for SEFOR than for a 1,000-megawatt fast-breeder
plant. As the quantity of accumulated fission products was much lower
in the SEFOR reactor, and since the amount of fission products contained in the reactor at any time determines the maximum amount
possible to be released in an accident, the fission product content of the
SEFOR reactor was of greatly reduced concern compared to a commercial plant.

But there were important hazards attending the operation of
SEFOR that are not present in the typical commercial nuclear power
plants operating then or now. To our knowledge, the siting of SEFOR
near Strickler, Arkansas, was unique in that it was the first nuclear reactor containing a substantial amount of plutonium *as fuel* (approaching
a half-ton) that had ever been sited anywhere in the free world except
on a government-controlled area intended to provide a secure "safe"
separation distance from the public. While located in a very low population density rural area in the Ozark Mountains, the SEFOR reactor
was on a relatively small (640-acre) site with minimal security provided
for control of access.

The "unique" hazards at SEFOR were two "special" materials used in the reactor:

- The primary fuel was the 239 isotope of plutonium, acknowledged to be the choice material for construction of nuclear fission weapons, and considered to be one of the most radiotoxic materials known should it gain access to the interior of the human body, particularly the lungs and bone tissue.
- The coolant used to remove the fission heat was liquid sodium, which is highly flammable if contacted with air or water. In addition, the sodium becomes radioactive as it cools the reactor, adding to the inventory of radioactive materials that could potentially be released. The radioactive sodium was primarily a disposal problem, rather than a primary hazard to the public, due to the limited reactor operation time and the containment provided.

The use of sodium as coolant poses challenging fire and explosion hazards and can complicate the design of the containment structure that forms the last defense against explosion events that could result in catastrophic release of radioactive materials to the environment. Nevertheless, this book focuses primarily on the hazards presented by the plutonium—because the plutonium fuel enriched to approximately 20% presented a new potential for an accidental *nuclear* explosion that might fail the final barrier to a catastrophic release of the fuel in aerosol (hot-particle) form to the atmosphere.

In order to simplify our task, we focus on the *additional* risks introduced if fast-neutron reactors replace thermal or slow-neutron reactors for generation of electric power. It is important to emphasize at the outset that the hazardous nature of both the extremely radiotoxic fissionable (fuel) materials and the fission products that inevitably accumulate during the operation of any nuclear fission reactor (slow or fast neutron) pose the primary hazards that differentiate the nuclear electric energy industry from the fossil fuel energy industry. We know that the potential consequences to the public of releases of large amounts of reactor fission-product contents, whether from thermal or fast-breeder reactors (of the same electric power generation capacity), pose similar risks to the public and the environment. This fact enables us to focus primarily on the *additional risks* associated with the adoption of fast-reactor technology.

So, the primary safety problem considered in this book becomes that of ensuring against very large amounts of *both* the nuclear fuel and accumulated fission products in a fast-neutron reactor being melted and vaporized—producing temperatures and pressures that could explosively breach the containment and allow catastrophic release into the atmosphere. Since the operation was of limited duration, the primary radioactivity hazards at SEFOR at the time of the December 1971 experiments considered here focus on the plutonium. Since we want to emphasize in this book the *additional risks* involving catastrophic releases of aerosolized fuel particles into the atmosphere, we first consider the baseline state of knowledge about the hazards of releases of radioactive *fission products* from reactors at the time the SEFOR experiments were completed.

The AEC-Acknowledged Public Hazards of Catastrophic Accidental Releases of Fission Products from Commercial Nuclear Reactors during the Period of SEFOR's Construction and Operation

Published estimates of damages that could result from large accidental releases of fission products were first prepared by experts for the government in 1957. The results were so threatening to the future of commercial power ventures being considered that the AEC exerted great pressure to try and minimize the concerns of the public. The AEC's task was difficult because the predicted hazards exceeded in magnitude that of any industrial (nonnatural) hazard that had been considered to that time. Ultimately, the AEC (and its successor government agencies) adopted methods of quantitative risk assessment (QRA) that rely on mathematical predictions of the likelihood of such accidents occurring to make the case that the risk could be "acceptable" in view of the positive benefits provided.

But there is a deep polarization of public opinion on such matters. In our opinion, the polarization is so profound that it is far beyond our ability to affect it in this book. We believe it is the most important question that must be answered by the regulatory agencies involved—the question of the acceptability of the risks to us all.

Our approach will be to briefly define the state of the argument that faces us as we consider the present proposals to jump-start the nuclear-electricity-generation industry using a new class of nuclear fast-fission reactors. As we do not want to attempt to address the stalemate that has developed in the public sector about the acceptability of the risks of nuclear power that could result from catastrophic accidental releases of radioactive fission-products from existing nuclear electric power plants, we will attempt here only to identify quasi-quantitatively the *additional* hazards, and the potential severity thereof, that could result from widespread adoption of fast-reactor technology. We believe those additional hazards are potentially game-changing.

To establish a baseline from which to compare the additional risks of fast-reactor technology, we begin with a brief description of the AEC's attempt in 1957 to inform the public of the hazards that could be realized from credible accidental releases of radioactive nuclear fission products to the atmosphere. In our opinion, the questions raised by WASH-740, based as it was on consideration by competent experts of the worst-case credible accidental release of such materials that might occur (in 1957), have never been satisfactorily addressed. Indeed, it appears that the study was never satisfactorily (or publicly) updated by the government to reflect the increase in size of the reactors that are in operation today compared to the reactors chosen for analysis in 1957.

WASH-740

Theoretical Possibilities and Consequences of Major Accidents in Large Nuclear Power Plants

Letter of Transmittal to Joint Committee on Atomic Energy

March 22, 1957.

Hon. Carl T. Durham,
Chairman Joint Committee on Atomic Energy,
Congress of the United States.

Dear Mr. Durham:

There is transmitted herewith a report of a study of the possible consequences in terms of injury to persons and damage to property, if certain hypothetical major accidents should occur in a typical large nuclear power reactor. More than two score leading experts in the sciences and engineering specialties participated in this study.

We are happy to report that the experts all agree that the chances that major accidents might occur are exceedingly small. This study constitutes a part of the commission's continuing effort on a broad front to understand and resolve this problem of possible reactor hazards so that we may proceed expanding atomic energy industry with full confidence that there will be few reactor accidents and that such as do occur will have only minor consequences. This effort and the work of translating the results into affirmative, concrete safeguards for protection of the public will, of course, be continued and expanded. Since the beginning of the reactor program the experts and the Congress and the public and the Commission have all been concerned with the causes of and the possible magnitude of damage from reactor accidents and with means of prevention. The subject was considered important enough to command four of the 60-odd sessions of the International Conference on the Peaceful Uses of Atomic Energy in Geneva eighteen months ago which, as you will recall, we initiated. One conference paper in particular gave estimates of the theoretical magnitude of damage. In May of last year, Dr. Libby presented to your Committee some estimations of the possible extent of harm and damage should a major accident occur.

This study has taken the form in which it is now presented to you as a means of responding to the Committee's specific request of last July 6. To produce such a study, it was necessary to stretch possibility far out toward its extreme limits. Some of the worst possible combinations of circumstances that might conceivably occur were included in the hypotheses in order that we might assess their consequences. The study must be regarded as a rough estimation of the consequences of unlikely though conceivable combinations of failure and error and weather conditions; it is not in any sense a prediction of any future conditions.

This has been a difficult study to make. There has fortunately been little reactor accident experience upon which to base estimates. Nuclear reactors have been operated since December 2, 1942, with a safety record far better than that of even the safest industry. More than 100 reactor years of regular operating experience have been accumulated, including experience with reactors of high power and large inventories of fission products, without a single personal injury and no significant depositions of radioactivity outside of the plant area. There have been a few accidents with experimental reactor installations as contrasted with the perfect record of safety of the regularly operating reactors. But even these accidents did not affect the public. This record which shows that safe operation can be achieved is due to skillful design, careful construction, and competent operation.

Looking to the future, the principle on which we have based our criteria for licensing nuclear power reactors is that we will require multiple lines of defense against accidents which might release fission products from the facility. Only by means of highly unlikely combinations of mechanical and human failures could such releases occur. Furthermore, the Government and industry are investing heavily in studies to learn more about the principles of safe reactor design and operation.

Framing even hypothetical circumstances under which harm and damage could occur and arriving at estimations of the theoretical extent of the consequences proved a complex task. To make the study we enlisted the services of a group of scientists and engineers of the Brookhaven National Laboratory and of another group of experts to serve as a steering committee. Through recent months these men have met with many additional expert advisors to offer judgment on the estimates arrived at. We are not aware of such a study having been undertaken for any other industry. We venture to say that if a similar study were to be made for certain other industries, with the same free rein to the imagination, we might be startled to learn what the consequences of conceivable major catastrophic accidents in those other industries could be in contrast with the action experience in those industries.

Remembering that this study analyzes theoretical possibilities and consequences of reactor accidents, we might note here the judgments presented on (1) possible consequences of major accidents and (2) the likelihood of occurrence of such major reactor accidents. The portion of the study dealing with consequences of theoretical accidents started with the assumption of a typical power reactor, of 500,000 kilowatts thermal power, in a characteristic power reactor location. Accidents were postulated to occur after 180 days of operation, when essentially full fission product inventories had been built up. Three types of accidents which could cause serious public damages were assumed. Pessimistic (higher hazard) values were chosen for numerical estimates of many of the uncertain factors influencing the final magnitude of the estimated damages. It is believed that these theoretical estimates are greater than the damage which would actually occur even in the unlikely event of such accidents.

For the three types of assumed accidents, the theoretical estimates indicated that personal damage might range from a lower limit of none injured or killed to an upper limit, in the worst case, of about 3,400 killed and about 43,000 injured.

Theoretical property damages ranged from a lower limit of about one half million dollars to an upper limit in the worst case of about seven billion dollars. This latter figure is largely due to assumed contamination of land with fission products.

Under adverse combinations of the conditions considered, it was estimated that people could be killed at distances up to 15 miles, and injured at distances of about 45 miles. Land contamination could extend for greater distances.

In the large majority of theoretical reactor accidents considered, the total assumed losses would not exceed a few hundred million dollars. (emphasis added)

As to the probabilities of major reactor accidents, some experts held that numerical estimates of a quantity so vague and uncertain as the likelihood of occurrence of major reactor accidents have no meaning. They declined to express their feeling about this probability in numbers. Others, though admitting similar uncertainty, nevertheless ventured to express their opinions in numerical terms. Estimations so expressed of the probability of reactor accidents having major effects on the public ranged from a chance of one in 100,000 to one in a billion per year for each large reactor. However, whether numerically expressed or not, there was no disagreement in the opinion that the probability of major reactor accidents is exceedingly low.

Some of the reasons for this belief follow:

First, industry and government are determined to maintain safety and protect the health and property of the public from nuclear hazards. The Congress has authorized and we in the Commission are carrying out a program of close and careful regulation and inspection. Thus the potential hazard of this new industry has been recognized in advance of its development and brought under a strict system of safety control before the occurrence of the incidents which in other fields have marked the birth of new industry and have subsequently led to control.

Secondly, the challenge of this new and important venture in man's application of the forces of nature has attracted able and energetic men into the work of assuring safe design and operation.

In the third place, multimillion-dollar efforts in research and development, both public and private, are directed toward identifying and solving safety problems. We know of no other industry where so much effort has been and is being spent on the definition and solution of safety problems.

Fourthly, the cost to the industry and government of reactor accidents, even of a minor nature, would be very high—much higher than for accidents in other industry. Self-interest, therefore, as well as public interest dictates avoidance of accidents.

To sum up, the report affirms that a major reactor accident is extremely unlikely. To reduce the matter of assumed hazards to comparative numbers, let us take the most pessimistic assumptions used and apply them to a case of 100 power reactors in operation in the United States.

Under these assumptions, the chances of a person being killed in any year by a reactor accident would be less than 1 in 50 million. By contrast, the present odds of being killed in any year by an automobile accident in the United States stand at about one in 5,000.

We are not surprised by the contents of the report, nor are we made complacent. The report serves to identify areas where continued research and development are needed, and areas where emphasis is needed in the further development of our regulatory program. It gives renewed emphasis to our belief that our research and development program and our regulatory program in the nuclear power field must continue with vigor to the end that the "conceivable" catastrophe shall never happen.

We would appreciate your regarding the attachment as an "advance" report. It is being reviewed for editorial and mechanical errors and omissions. Copies of the report as corrected will be furnished to you at an early date.

Sincerely yours,

(Signed) Harold S. Vance,
Acting Chairman

WASH-740 Was Not Updated

The results of WASH-740, prepared by a blue-ribbon panel of scientists from the Brookhaven National Laboratory, delivered an unsettling picture to the public about the risks of embarking on a program to develop a large number of fission reactors for generating electric power. The predictions of death, injury, and possible abandonment of large tracts of land downwind of such accidents appeared to exceed any hazards expected previously from industrial enterprises.

WASH-740 was never updated to reflect the changes that would be required when much larger reactors were being suggested by the 1970s. Then in 1976, Dr. Richard E. Webb, who had testified in 1972 about the nuclear explosion hazards of fast reactors before the Joint Committee on Atomic Energy of the U.S. Congress (see Notes), published the book *The Accident Hazards of Nuclear Power Plants*. In the book, Dr. Webb updated the predictions of WASH-740 to account for the size of reactors planned in 1976, nominally 1,000 megawatts, from the reactor size assumed in WASH-740, ~200 megawatts (electric). Dr. Webb assumed that the damages would scale directly with the total fission rate, which would be approximately six times larger in 1970 than in 1957. Quoting Webb:

> To estimate the maximum sequences of any reactor accident, we adjust the estimates of the maximum possible reactor accident as given in the 1957 Atomic Energy Commission (AEC) report, Theoretical Possibilities and Consequences of Major Accidents in Large Nuclear Power Plants (WASH-740), to account for the six-fold increase in the highly intense, short-lived radioactivity

and the fifteen-fold increase in the long-lived radioactivity in present day reactors. The maximum conceivable consequences of the worst accident are as follows: (1) a lethal cloud of radiation with a range of seventy-five miles and a width of one mile; (2) evacuation or severe living restrictions for a land area the size of Illinois, Indiana, and Ohio combined (120,000 square miles), lasting a year or possibly longer; and severe long-term restrictions on agriculture due to strontium 90 fallout over a land area of the size of about one half of the land east of the Mississippi River (500,000 square miles), lasting one to several years, with dairying prohibited "for a very long time" over a 150,000 square mile area. There are other consequences not here estimated for water-cooled reactors, such as genetic damage. *The potential accident consequences for the LMFBR— especially with respect to plutonium contamination, which may be a gravely serious lung-cancer hazard—will be discussed later, since they will depend on the explosion hazard unique to that reactor.* (emphasis added)

Plutonium Fuel

The bulk of the radioactive materials contained in any operating reactor comprises the fuel and its fission products. SEFOR was fueled with plutonium, the core enriched to roughly 20% plutonium with the balance primarily uranium 238. Because of the relatively low total production of fission products in the SEFOR reactor during its operating period (compared to a commercial reactor), the focus of the accident hazard potential for SEFOR was the plutonium. So what is the special concern about plutonium?

Typical commercial reactors operating in the 1960s "burned" U_{235}. Further, all of these reactors were (and remain) "thermal," or slow-neutron reactors, with average neutron speeds around 2,000 meters per second (around 4,500 miles per hour). The exceptions during that time period (1950s and '60s) were SEFOR and a commercial, though "demonstration," fast-neutron "breeder" reactor known as FERMI 1, which was constructed on the shore of Lake Erie about 30 miles from Detroit. FERMI 1 was fueled with enriched U_{235} metal, whereas SEFOR was fueled with enriched Pu_{239} oxide. FERMI 1 and SEFOR were fast-neutron reactors, with fission neutrons driving the chain reactions with average neutron speeds typically about 20 *million* meters per second (around 45 *million* miles per hour). Importantly, plutonium oxide, in contrast to plutonium or uranium metal, had been shown to exhibit a much larger negative Doppler coefficient. It was the magnitude of this negative Doppler coefficient that constituted the safety margin that was to be demonstrated at SEFOR.

Hence, SEFOR became, in the mid-1960s, the first (relatively) large *fast-neutron* reactor fueled with plutonium to be built in the United States. The SEFOR reactor was the first (large) reactor to couple the potential for very fast-neutron reaction rates (more challenging to control than the slow-neutron "thermal" reactors) with the use of plutonium as fuel. Because of the perceived high hazard of plutonium should it be released to the atmosphere, and the increased risk of *nuclear explosion* energy release that accompanies the operation of fast reactors, concerns of knowledgeable scientists, medical doctors, and industry/government parties were very real.

The general population of northwest Arkansas and surrounding states appears to have been largely unaware of such concerns, instead receiving assurances by the AEC and the nuclear industry that the reactor would safely "demonstrate" that widespread fears of a nuclear explosion of magnitude to allow significant amounts of the plutonium in the reactor to enter the atmosphere were unfounded for the fast reactors then planned to be fueled with plutonium oxide (not metallic plutonium). It didn't hurt the government's *promotional* case that the proposed commercial fast-breeder reactors (nearly a thousand were tentatively planned), all burning oxides (oxygenated forms of the plutonium), were widely viewed as the best, if not the only, potential solution to the country's energy problems. The nuclear explosion question was now going to be addressed with the use of plutonium oxides as fuel. How well it would be addressed is the principal concern of this book.

Important Additional Hazards of Plutonium

Plutonium began appearing in measurable amounts after about 1940 as a result of research driven by the development of nuclear fission weapons. It soon became the fissile (fissionable) material of choice for nuclear weapons production. The first microgram quantity of plutonium was produced at the University of California in 1941. Within a few years, plutonium production reactors were operating on the Hanford Reservation spread out along the Columbia River in Washington State. The plutonium produced in those so-called "production" reactors was chemically separated into near pure form in purpose-built separation plants at Hanford.

The hazards of plutonium are due to its chemical (heavy metal) toxicity and its radioactivity. Plutonium is considered one of the most radiotoxic materials known. The first bomb-grade plutonium at Hanford was assembled as a hollow sphere approximately 4 inches in diameter weighing about 6 kilograms (about 14 pounds), transported to the Trinity site in New Mexico and exploded in August 1945. A similar plutonium sphere was used in the bomb that was dropped on Nagasaki. In the Trinity test and in the Nagasaki bomb, only about one-fifth of the core, approximately 1 kilogram, fissioned, and approximately 1 gram (1/30th of an ounce) of its mass was converted into explosive energy equivalent to ~21,000 tons of TNT.

Most importantly to the population around the SEFOR site, the reactor's enriched plutonium fuel made it a fast reactor, with potential neutron speeds tens of thousands of times faster than occurs in thermal reactors. Such neutron speeds results in much higher fission reactivity rates, which makes the engineering design for reactor control much more challenging. The result is a heightened potential for accidental explosions; enrichment to the 20% level is an important step along the path required to assemble a nuclear weapon with very high explosion yield— witness the present heightened concerns about the dangers of fissile fuel enrichment by countries potentially unfriendly to the United States.

We will very soon focus on the explosion potential of fast reactors. However, as we are primarily concerned about the potential for radioactive substances accidentally released from reactors to enter the air (enabling *inhalation* by humans), we will first briefly consider what we think we know, and in some considerable measure has been verified, about *quantifying* the potential for humans to inhale air containing radioactive materials and the damage that can result. Of course, the libraries are filled with books containing such information, but it is so extensive and complicated that there is little likelihood of the general population achieving a satisfactorily complete understanding of that scientific knowledge to be able to collectively agree on the dangers involved and the measures that can be taken to mitigate the hazards. Instead, the public continues to argue, largely utilizing various forms of wishful thinking, that the hazards to humankind of adopting such technological measures as fast plutonium-fueled reactors to generate electricity are overblown, in much the same way as the general population seems split on the dangers of global warming. It is ironic that the present argument for the "nuclear" solution, involving considerable uncertainty, and so presciently negated with the occurrence of nuclear accidents "that couldn't or wouldn't happen" (like Chernobyl and Fukushima), is being suggested to solve the problem of global climate change. We believe that either of these threats, in time, could "do us in"; hence we must get this argument right lest we create even more problems that we cannot solve. Our first step is to set down briefly what the concerns are for the radioactivity hazards that could attend large-scale plutonium fast-reactor operation to generate electricity.

While plutonium is "new," radioactivity has been with us all along. It is common knowledge that radioactive materials are pervasive throughout the earth and its oceans and atmosphere; indeed, radioactive materials, and the attendant hazards, are accepted as pervading the entire universe. *But there has been an important change during the last century*, beginning about 1940—we have managed to rearrange the materials composing the earth so as to concentrate them in more compact form for our (presumed) benefit. We have learned how to "assemble" or "concentrate" the unstable elements like uranium and heavier elements (like plutonium) in configurations with which we can design systems to release the powerful forces that hold the atoms of these materials together. The splitting of the nuclei of such atoms, which requires their assembly in greatly purified form, releases energy in the form of heat at rates thousands of times greater than we could accomplish before we "discovered" the mechanism of chain reaction nuclear fission.

The "assembly" discovery led to our building weapons that produce extraordinary explosion power utilizing *uncontrolled* fission processes, but the technology involved was soon extended to build nuclear power plants to provide *controlled* release of fission-generated heat energy to drive electricity generators. We now are well into an age of crippling fear of the potential for accidents in nuclear power stations to suffer releases of radioactive materials to the environment with catastrophic consequences extending from the release site to very large distances—potentially involving large segments of the human, plant, and animal populations. The amount of radioactivity resulting from nuclear electric power production in the world already may exceed the amount of radioactive materials that could be released if all of the atomic weapons existing were exploded (this amount is thought to be classified), and it is expected to grow even further, with few solutions for satisfactorily safe disposal in sight. Such is the present state of the nuclear waste disposal dilemma. The problem is at once simple to state and extraordinarily difficult to solve. The radioactive fission products that build up in fission reactors must be contained safely within the plants until they are required to be removed (periodically, for refueling). When they are removed, they must be placed where they cannot escape until the radioactive decay processes proceed to the point where the radioactivity hazard "disappears" (which it will eventually do). The problem is that the time period for radioactive decay processes to reach safe levels ranges, depending on the individual species (of which there are hundreds), from time periods of a few minutes (or small parts thereof) to thousands of years. Humankind has thus far been unable to solve this problem, which appears to be intractable, perhaps more for political/sociological reasons than scientific ones.

Fast-Reactor Explosion Hazards

This book's principal focus is the potential for accidental explosions occurring in fast nuclear reactors that could result in the release of large amounts of radioactive materials, including plutonium or other fissile isotopes of trans-uranium materials, into the environment. Aside from the releases of natural radioactive materials that occur in nuclear (electric) power generation due to mining and preparation of nuclear fuel and some arguably minor but necessary "planned" releases of radioactive gases produced in operating reactors, the potential for radioactive materials contaminating the earth as a result of nuclear-powered electricity generation rests in large part in our lack of inability to ensure against releases due to accidental or intentional (terrorist) breaching of the reactors and their containment structures. In light of Chernobyl and Fukushima, as well as several other potentially disastrous near misses, we have a long way to go in solving this problem.

Increasing fears of "radioactivity" were spurred by the results of the atomic bomb attacks on Japan, where the direct results of exposure to significant amounts of radiation were extreme, injuring and killing people in huge numbers within days or weeks. Following WWII, the increased releases of radioactive materials—especially associated with open-air weapons testing that resulted in exposure to the public to great distances by winds carrying the airborne radioactive gases and liquids and solids small enough to remain suspended in the atmosphere—resulted in the fallout problem. We now know that a large part of the radioactive materials in the "clouds" produced by nuclear weapons remain suspended in the atmosphere for sufficient time to allow some of the material to become effectively diffused throughout the earth's atmosphere. It is by no means uniformly distributed, but we can be sure that it has, or will, spread throughout the atmosphere, and indeed, in sufficient time, the entire earth. The part that does not stay suspended falls out onto the earth, constituting additional routes by which the radioactive materials can contact humans. Finally, such radioactive materials, which by definition are unstable, change to other materials via a process called radioactive decay. The so-called "daughter" products may also be radioactive, which means that all of the materials so formed will ultimately further decay into products until materials that are stable—that is, not radioactive—result.

In the case of electric-power-generating reactors, the radioactive products of the fission process and the subsequent decay processes that inevitably occur could theoretically be contained so that humanity is never exposed to significant amounts. Significant amounts are generally considered to be amounts that would not exceed those already present on earth. What "significantly" means is, like the nuclear explosion potential of fast reactors, at best argumentative.

Explosion Containment

Research for this book showed that at the time SEFOR was constructed:

- The maximum (economical) containment that could be provided to ensure that a major release of aerosolized fuel and fission products from a commercial-size LMFBR could not occur was estimated to be about 1,000 pounds (1/2 ton) of TNT explosive equivalent. Provision of significantly more robust containment appeared, on its own, to make the plants economically uncompetitive.

- The maximum possible (nuclear) explosion yield that could occur in a fast reactor due to accident or natural disaster causes was then argumentative, at best, and had been predicted by reputed scientists to exceed by an order of magnitude (at least ten times) the explosion energy yield that could be economically provided for.

Arkansas Gazette, May 15, 1964
Arkansas congressmen and businessmen pose with officials of the AEC at a signing of a contract for a multimillion-dollar atomic reactor project near Fayetteville. At right, seated, is Glenn Seaborg, chairman of the AEC.

SEFOR, operational, about 1970, Strickler, Arkansas.

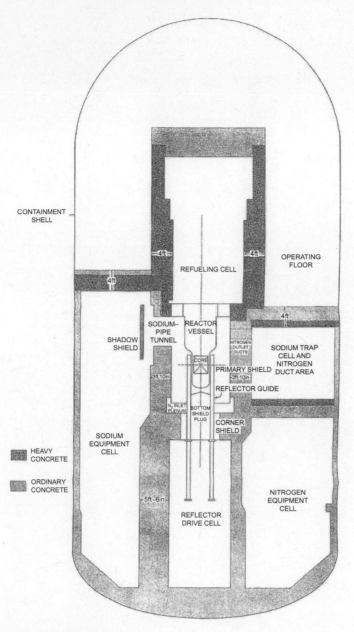

CONTAINMENT
SHELL

4ft

4ft

4ft

REFUELING CELL

OPERATING
FLOOR

4ft

SHADOW
SHIELD

SODIUM–
PIPE
TUNNEL

REACTOR
VESSEL

NITROGEN
OUTLET
DUCTS

SODIUM TRAP
CELL AND
NITROGEN
DUCT AREA

CORE

3ft-10in

PRIMARY SHIELD

3ft-10in

REFLECTOR GUIDE

N₂ INLET
PLENUM

BOTTOM
SHIELD
PLUG

CORNER
SHIELD

SODIUM
EQUIPMENT
CELL

HEAVY
CONCRETE

ORDINARY
CONCRETE

5ft-6in

NITROGEN
EQUIPMENT
CELL

REFLECTOR
DRIVE CELL

Containment vessel
Height 115 feet – Diameter 50 feet
Reactor vessel near center.

View through containment door. Grappling hook over reactor in background. Post-closure.

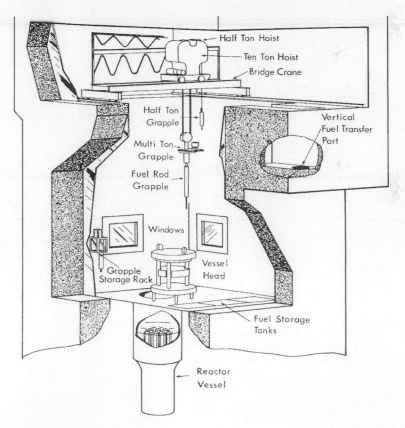

Refueling cell atop reactor.

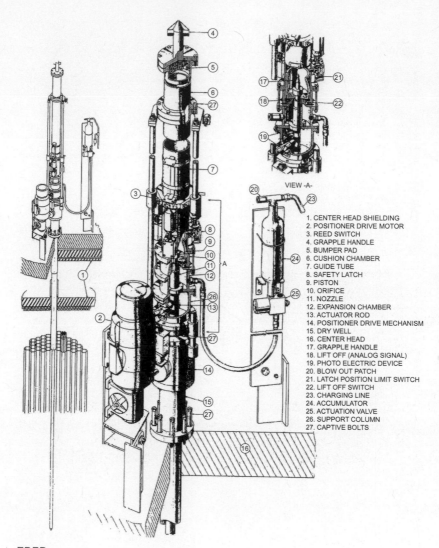

VIEW -A-

1. CENTER HEAD SHIELDING
2. POSITIONER DRIVE MOTOR
3. REED SWITCH
4. GRAPPLE HANDLE
5. BUMPER PAD
6. CUSHION CHAMBER
7. GUIDE TUBE
8. SAFETY LATCH
9. PISTON
10. ORIFICE
11. NOZZLE
12. EXPANSION CHAMBER
13. ACTUATOR ROD
14. POSITIONER DRIVE MECHANISM
15. DRY WELL
16. CENTER HEAD
17. GRAPPLE HANDLE
18. LIFT OFF (ANALOG SIGNAL)
19. PHOTO ELECTRIC DEVICE
20. BLOW OUT PATCH
21. LATCH POSITION LIMIT SWITCH
22. LIFT OFF SWITCH
23. CHARGING LINE
24. ACCUMULATOR
25. ACTUATION VALVE
26. SUPPORT COLUMN
27. CAPTIVE BOLTS

FRED

Fast Reactivity Excursion Device

FRED was an electromechanical device designed to accurately change the vertical position of a poison rod in the center of the reactor core. When the poison rod was centered in the core, it provided zero positive reactivity. Reactivity was added by pushing the rod out of the core at an accurately controlled rate, to an accurately specified position, after which the rod fell back into the core, terminating the reactivity addition. The time from start of rod movement to the chosen extent (larger extent gave increased reactivity addition) was intended in the tests described to be 0.1 *second*.

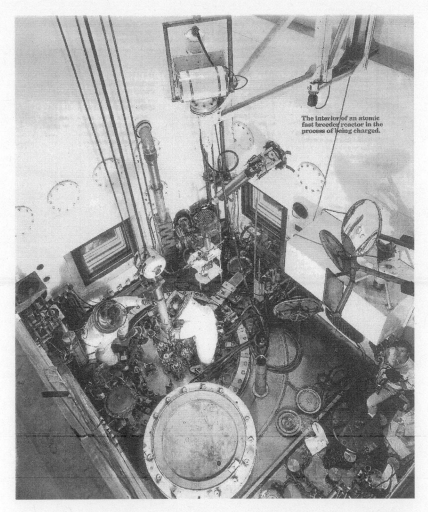

The interior of an atomic
fast breeder reactor in the
process of being charged.

This photo was discovered in a magazine entitled *The Russian Disaster: A Survival Handbook for the Nuclear Age*, by Bernard Crossfield, PhD, published following the Chernobyl, Ukraine, disaster by Paladin Press. The text that had been added is incorrect; it is not a breeder reactor, nor is it being charged with fuel. At the time of the photo it was charged with approximately 900 pounds of plutonium fuel. The photo was taken from overhead the SEFOR reactor in the refueling cell. The reactor top has been removed, and the two men suited appear to be installing the FRED reactivity device. If the men pictured are recognized in this picture, we would appreciate a contact with the authors.

This photograph, an enlargement of the center section of the previous photo, appears to show the Fast Reactivity Excursion Device being installed in SEFOR. The FRED was installed in the center of the reactor.

Arkansas Gazette, March 7, 1967
J. Robert Welch (center), president of Southwestern Electric Power Company at Shreveport and president of Southwest Atomic Energy Associates, which is sponsoring construction of the SEFOR Reactor near Fayetteville, discusses the project with Dr. Bert Wolfe, manager of SEFOR Engineering and Development (left) and Dr. Karl Cohen, manager of General Electric's Advanced Products Division, designer and builder of the fast-oxide, sodium-cooled reactor of which Dr. Cohen is considered the originator.

This picture of the abandoned reactor, in a hayfield overgrown with trees, was taken from adjacent Arkansas Hwy 265, probably in the period 2000–2010.

2019. The above-ground top half of the steel containment cylinder has been removed, and the reactor is being extracted.

2019. Installed in a secure container, leaving the site headed to a Nevada disposal site—after fifty years.

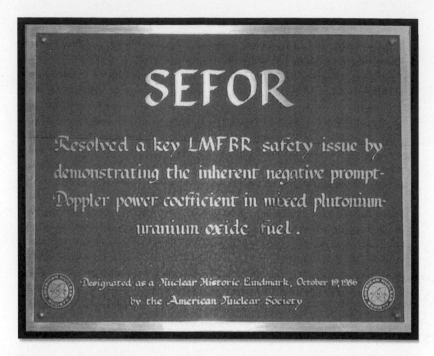

SEFOR demonstrated the Doppler effect for nuclear safety of fast-breeder reactors fueled with plutonium-oxide fuel. Despite the acknowledged importance thereof, this book shows that the results of the SEFOR experiments did not carry the message of guaranteed safety that the public was led to believe.

CHAPTER 5

Nuclear Explosion Potential in Fast Reactors

For the fast breeder to work in its steady-state breeding condition you probably need something like a half ton of plutonium. In order that it should work economically in a sufficiently big power-producing unit, it probably needs quite a bit more than one ton of plutonium. I do not like the hazard involved. I suggested that nuclear reactors are a blessing because they are clean. They are clean as long as they function as planned, but if they malfunction in a massive manner, which can happen in principle, they can release enough fission products to kill a tremendous number of people. . . . If you put together two tons of plutonium in a breeder, one tenth of one percent of this material could become critical. I have listened to hundreds of analyses of what course a nuclear accident can take. Although I believe it is possible to analyze the immediate consequences of an accident, I do not believe it is possible to analyze and foresee the secondary consequences. In an accident involving a plutonium reactor, a couple of tons of plutonium can melt. I don't think anybody can foresee where one or two or five percent of this plutonium will find itself, and how it will get mixed with some other material. A small fraction of the original charge can become a great hazard.

Edward Teller, 1908–2003

In September 1972, Dr. Richard E. Webb appeared before the Joint Committee on Atomic Energy of the U.S. Congress in opposition to the construction of the Clinch River Breeder Reactor Project (CRBRP). Dr. Webb testified to the potential danger of nuclear explosions in fast reactors that could be severe enough to compromise the containments then proposed to prevent catastrophic releases of radioactive materials to the environment. Webb's testimony included statements that the AEC was taking a chance with public safety by approving SEFOR as well as the CRBRP before sufficient research had been done to determine the maximum explosion potential that was possible in liquid-metal-cooled fast reactors. The written record of his testimony was followed by the AEC's response, which we believe was unjustifiably dismissive. Webb prepared a written rebuttal to the AEC's response and delivered it to the AEC in July 1973. Webb's rebuttal was not made public by the AEC (to our knowledge), but we know the AEC received his rebuttal remarks, as the University of Arkansas Library determined that a copy of the rebuttal document is held by the Nuclear Regulatory Commission (NRC) in Washington, D.C.

We obtained a copy of Webb's rebuttal remarks to the AEC from Purdue University, whose library holds Dr. Webb's papers relating to nuclear safety, in 2014. After careful review, we concluded that the rebuttal remarks, which exceed two hundred pages, constitutes one of the best discussions of fast nuclear reactor safety explosion hazard potential (that could be considered understandable by the public) that was available at the time it was written—mid-1973, one year after SEFOR closed. Webb's rebuttal "report" is particularly significant because it can be confirmed that the AEC (now NRC) received it (presumably in 1973). The rebuttal remains today an excellent example of independent-expert testimony on the specific subject of fast-reactor explosion potential.

We present here selections pertaining to reactor safety from Dr. Webb's testimony to Congress in 1972 followed by selected sections from the AEC's response. We then present selected material from Webb's rebuttal, which we believe remains insufficiently considered to this day. Finally, a stand-alone section from Webb's rebuttal entitled "Basic Theory of LMFBR Nuclear Runaway in More Detail" is presented. This section provides excellent information for understanding the limits of nuclear explosion potential in fast reactors. While Webb's testimony and his rebuttal will be heavy sledding for the typical reader, we think it important to make it a matter of public record in this book. In our opinion, Webb's presentations, particularly the technical discussions about the potential for nuclear explosions in fast reactors, remain the most accurate and sober description of such risks available to the lay reader. Indeed, it is suitable for careful study as groundwork for students as well as practitioners of nuclear engineering.

It is our hope that Webb's statements to Congress and the review thereof by the AEC in 1972, along with Webb's rebuttal in 1973, and the effective dismissal by the AEC of essentially all of the points he raised less than a year before the AEC was dissolved, will be considered appropriate for serious consideration by any party desiring to know more about this critically important subject of nuclear fast-reactor safety—including the public and students at advanced levels in related engineering and science disciplines.

The AEC was abolished in 1974. In this chapter, we consider the relevance of Webb's submissions to the AEC regarding the fast-reactor safety debate. Dr. Webb's statement to Congress and his rebuttal remarks to the AEC appear below.

Excerpt from Hearings before the Joint Committee on Atomic Energy, Congress of the U.S., 92 Congress, Second Session, Sept. 1972: Liquid Metal Fast Breeder Reactor (LMFBR) Demonstration Plant

Excerpts from Statements by Dr. Richard E. Webb

Bloomington, Ind., September 20, 1972.

John O. Pastore
Chairman, Joint Committee on Atomic Energy,
U.S. Senate, Washington, D.C.

Dear Senator Pastore:

I am enclosing my statement concerning the Liquid Metal Cooled, Fast Breeder Reactor Demonstration Plant. Please accept it for inclusion in the record of your hearings on the LMFBR Demo.

My background and expertise briefly is as follows:

1. Ph.D. in Nuclear Engineering, the Ohio State University, March 1972. My Ph.D. thesis concerns the explosion potential of the LMFBR.

2. Served four years (1963–1967) with the AEC's Division of Naval Reactors, during which my primary responsibility was for the nuclear reactor portion of the Shippingport Pressurized Water Reactor.
   a. Certificate of successful completion, Bettis Engineering School of the AEC's Bettis Atomic Power Laboratory (1965)
   b. Reactor Plant Training (one month) at the Navy's DIG Prototype Reactor Plant at the AEC's Knolls Atomic Power Laboratory (1966).
3. Worked one-half year at Big Rock Point Nuclear Power Station (Boiling Water Reactor) at Charlevoix, Michigan as associate engineer with reactor engineering duties in 1967. (I was offered a position with the LMFBR Program Planning Office in 1968.)
4. B.S. Engineering in Physics, University of Toledo, 1962.
5. Presently preparing a book on criteria and procedure for establishing sound public decision with respect to civilian nuclear power at Indiana University's School of Public and Environmental Affairs (Science, Technology and Public Policy section).

Sincerely yours,

Richard E. Webb

Enclosure: Statement on the LMFBR Demonstration Plant.

LMFBR Demonstration Plant

(Statement by Richard E. Webb)

Summary

The Liquid Metal Fast Breeder Reactor Demonstration Plant (LMFBR Demo) should not be built (not now at least) because the maximum explosion potential has not been scientifically determined. Because the LMFBR Demo will contain up to 1.3 tons of Plutonium and a large amount of fission product radioactivity, which absolutely must not be allowed to be spewed into the environment by a reactor plant explosion, the unknown explosion potential of the LMFBR Demo makes it imperative that the present plans for constructing and operating such a reactor be discarded in favor of further, more thorough, theoretical and experimental research into the said explosion potential. . . .

Explosion Potential

The "Environmental Statement" (WASH-1509, April 1972) issued by the Atomic Energy Commission for the LMFBR Demo states that the substantiation of the claim that the reactor will be safe must await the issuance of the preliminary Safety Analysis Report (PSAR) when the construction permit application is filed with the AEC. (See E.S., p. 37, 107). But it is obvious from reading the Environmental Statement that the AEC and the Joint Committee on Atomic Energy have prejudged the question of safety. For example, Congress has already authorized the LMFBR Demo and appropriated the money for it (E.S., p. 1). Furthermore, the AEC asserted in the Environmental Statement that the provisions in the reactor containment structure for "blast and missile protection within the inner barrier provide substantial margins against major potential energy release for *all* classes of accidents" (E.S., p. 54; emphasis added). The AEC added: "While it is impossible to postulate with precision the detailed course of accidents, including their likelihood and possible environmental consequences, *it is possible to place bounds on such accidents*" (E.S., p. 119; emphasis added).

These statements have no scientific foundation. Based on my knowledge of the state of the science of LMFBR explosion calculations, there is no chance that the aforesaid PSAR will substantiate such conclusions. Therefore, the construction of the LMFBR Demo should not be undertaken until after the necessary theoretical and experimental research is conducted, if such research demonstrates safety. The alternative is for Congress to recall the authorization and appropriation for the LMFBR Demo, wait for the issuance of the PSAR, and its review by the AEC and the Public, then hold public hearings on the safety of the LMFBR Demo.

The basis for my assertion is contained in my Ph.D. dissertation (thesis) which was submitted to, and approved by, the faculty authorities in the department of Nuclear Engineering at the Ohio State University. The title of the dissertation is "Some Autocatalytic Effects during Explosive Power Transients in Liquid Metal Cooled, Fast Breeder, Nuclear Power Reactors (LMFBRs)," the Ohio State University (1971).[1] A copy of the dissertation was sent to the director of the AEC's Division of Reactor Licensing (Mr. Peter Morris), which the Committee could borrow.

To summarize the conclusions of my dissertation, the calculational methods for determining the maximum explosion possible in an LMFBR have not been developed to include all possibilities, and their combinations, for autocatalytic phenomena during and after an initial nuclear runaway. That is, there are conceivable mechanisms by which "reactivity" can or might be rapidly "inserted" due to the motion of fuel material resulting from an initial core explosion or meltdown event. (Recall that in fast reactors, a core meltdown presents a mechanism by which reactivity can increase semi-rapidly and trigger disruptive or explosive power pulses.)[2] In other words, an initial event, or series of events, might cause the reactor to feed itself a massive dose of "reactivity" which would amplify the initial runaway, or cause a very severe secondary runaway; either of which might lead to a disastrous explosion.

When the calculational methods are developed to include all possible autocatalytic effects, they would still need experimental confirmation. Moreover, as I asserted in my thesis (p. 44), the present calculation methods "have not been confirmed experimentally for power reactor designs". For example, it has been claimed by Hirakawa and Klickman[3] that the KIWI-TNT power excursion experiment (TNT stands for Transient Nuclear Tests) has confirmed the MARS fast reactor excursion computer code. (The basic theory in MARS is the Bethe-Tait theory, which is partially used in the more advanced explosion codes such as VENUS. This theory provides the reactivity feedback mechanism that ends or "shuts down" the power excursion, and thereby, limits the explosion force.) However, though the post facto MARS calculation of energy yield agreed fairly well with the KIWI-TNT measurement, the power pulse height (peak power), pulse shape, and pulse width as calculated by the MARS code are completely different than the KIWI-TNT experimental results. I used a simple thermal expansion model which excludes the basic theory in MARS that was thought to be tested (i.e., the Bethe-Tait theory), and calculated all four of the above items in excellent agreement with the experimental results.[4] This strongly indicates that the inherent shutdown reactivity mechanism in the KIWI-TNT experiment was not the Bethe-Tait mechanism, but one due to the simple thermal expansion of the KIWI core; and that agreement between the MARS value of energy releases and experimental measurement was coincidental. In support of my conclusion, Jankus stated that the "Bethe-Tait assumption is definitely unjustified" for the KIWI-TNT excursion.[5] Furthermore, KIWI was not a fast reactor. Therefore, the KIWI-TNT explosion test has not been shown to be a confirmation of LMFBR explosion theories.

The SEFOR power excursion tests, which were performed to confirm the mitigating action of the Doppler effect for fast reactors, cannot be considered as proving out the LMFBR explosion calculational methods because the SEFOR excursions were not designed to lead to an explosion.[6] The tests involved (1) relatively mild rates of programmed reactivity insertion, (and then the total reactivity inserted was limited to a small amount); (2) designed Doppler feedback magnitudes that were much greater than typical 1000 MWe LMFBR design values; and (3) automatic termination of the power transient by control rod scram (probably preprogrammed) to ensure against unexpected secondary excursions. Because of the strong Doppler and the limited amount of total reactivity that was inserted, the strongest power excursion tested was easily stopped with only about a 10% rise in the fuel temperature, which means that the SEFOR tests approached no threshold for meltdown or explosion. Normally in LMFBR accident calculations one assumes that the initial reactivity insertion is not limited, but is unrelenting. Thus in a real accident situation the Doppler effect alone would not be sufficient to terminate the power excursion, and the core would continue to generate energy until there is an explosive or disruptive "disassembly" of the core that finally stops the power excursion and shuts down the reactor, if one could still call a reactor destroyed a "reactor." (Just how severe the explosion is and whether aggravated by autocatalytic effects is my main concern.)

Therefore, although the SEFOR tests were very useful in demonstrating the Doppler mitigating mechanism, and were evidently successful in that regard they provide no confirmation of explosion calculational methods. This is just as well, since there is a report which indicates that SEFOR was not designed to contain severe explosions.[7] With one-half ton of Plutonium in the SEFOR reactor, it appears that the AEC simply took a chance with the public safety by purposely causing power excursions, which one tries normally to prevent in power reactors, to test a safety effect (Doppler feedback) that was not beforehand demonstrated in a fast reactor power excursion. (SEFOR is now being decommissioned now that the tests are finished.) Whereas, prudence would suggest that such tests involving so much Plutonium should have been conducted only after a thorough research into auto-catalytic reactivity effects was completed to establish the maximum possible accident. Then prudence would suggest that such a test reactor would be placed deeply underground just in case something was over-looked. (The EBR-I, BORAX-I, and SPERT-I reactors all suffered accidents because the power excursions were under-calculated.[8] But instead, SEFOR was built above ground and may have been without explosion containment. Similarly, the LMFBR Demo would be an experiment with unknowns, involving 1.3 tons of Plutonium, and fission product Strontium-90 and Cesium-137 and the like. That is, the LMFBR Demo is simply a chance that will be taken with the health and safety of the Public if allowed to be built without a firm ground of scientific research to establish the containment design.

I mentioned so far the lack of experimental confirmation of existing calculational methods, as well as the inadequacy of the calculational methods from the standpoint of autocatalytic reactivity effects. The improved calculational methods for predicting the LMFBR explosion potential, once developed, would still require experimental confirmation, just as was done to some extent for the Doppler effect in the SEFOR tests. To be sure, fast reactor explosion tests were proposed by Nims at the 1963 Argonne National Laboratory Conference on "Breeding, Economics and Safety in Large Fast Power Reactors."[9] Nims considered the straightforward approach of simply building a prototype reactor, causing the core to meltdown, and observing the resulting explosion. Such tests would have to be repeated in a variety of ways in an effort to cover all possible or conceivable ways in which the core might meltdown. Nims indicated that the costs for such a series of tests would be prohibitive, since a series of costly reactors would have to be built, just to be destroyed. As an alternative he proposed a series of partial core meltdown experiments, short of explosion, to learn the manner in which the core would meltdown; and then with a more confident understanding of core meltdown acquired by such tests, full scale reactor meltdown tests would be designed and performed to determine the severity of the explosions associated with the prior established core meltdown patterns.

Nims argued that this alternate scheme may provide the desired information regarding LMFBR explosion potential at acceptable cost. I would add that the development of improved calculational methods regarding autocatalytic effects, that I contend is necessary, would be of help in designing such explosion experiments. (Of course, there is the possibility that such improved calculational methods might predict with confidence that the explosion potential of LMFBRs is simply too great to ever consider building LMFBRs at all.) The LMFBR Program Plan (Volume 10, Safety) provides for studies of the necessity for such explosion testing.[10] (The Plan has adopted the alternate scheme investigated by Nims as that which is to be considered, without mentioning the more direct method of testing prototype reactors.) I have seen no results of such studies. Presumably, they are still being conducted. But regardless of their outcome, until improved theoretical methods are developed and tested by reactor explosion experiments, claims that the LMFBR containment structure is designed to contain "all classes of accidents" and that "it is possible to place bounds on such accidents" will continue to be groundless. Accordingly, if the United States is to pursue LMFBR development, we should discard the plans for a demonstration power reactor in favor of further research terminating in explosion testing, unless the theoretical research proves that LMFBRs are inherently unsafe, so that we can be assured of confining the Plutonium and other radioactivity in the event of the worst possible LMFBR accident.

(The foregoing material was submitted to the AEC for comment Correspondence and comment follow:)

September 25, 1972.

Mr. Robert E. Hollingsworth
General Manager, U.S. Atomic Energy Commission,
Washington, D.C.

Dear Mr. Hollingsworth:

Enclosed is a "Statement on the Liquid Metal Cooled, Fast Breeder Reactor Demonstration Plant" by Richard E. Webb, Ph.D. The Committee is considering the inclusion of this statement in the public hearing record on the arrangements for construction and operation of the demonstration liquid metal fast breeder reactor. Please review the enclosed document and supply the Committee with the Commission's comments on it.

Sincerely yours,

Edward J. Bauser
Executive Director

Atomic Energy Commission
Washington, D.C., October 25, 1972.

Mr. Edward J. Bauser
Executive Director, Joint Committee on Atomic Energy,
Congress of the United States

Dear Mr. Bauser:

In accordance with the request in your letter of September 25, 1972, enclosed is the AEC staff Review of a "Statement on the Liquid Metal Cooled, Fast Breeder Reactor Demonstration Plant" by Richard E. Webb, Ph.D.

In its comments the staff addresses mainly Dr. Webb's views on breeder reactor safety. . . . Our review indicates that from technical and legal standpoints the Statement offers no justification for reversing the AEC's current plans for designing, constructing and operating the LMFBR Demonstration Plant.

If we can provide you with any additional information in this regard, please do not hesitate to contact us.

Sincerely,

John O. Erlewine,
Deputy General Manager

AEC Staff Review of Dr. R. E. Webb's Statement on the LMFBR Demonstration Plant

Safety Issues Pertinent to the LMFBR Demonstration Plant

The Division of Reactor Development and Technology has under way an extensive base technology and development programs for the purpose of providing engineering and safety understanding and thus assuring the success of the LMFBR program objectives, including the Demonstration Plant. Volume 10 of the LMFBR Program[11] covers all questions relating to the LMFBR Safety program and in particular such questions as raised by Dr. Webb, which fall in the category of hypothetical accidents and their consequences. In the area of hypothetical accidents, the safety program has as its objective the understanding of phenomena related to hypothetical events and their consequences through the conduct of extensive in-pile and out-of-pile testing as well as analytical programs which complement the experiments. This understanding will provide realistic bounds and estimates of risk so as to permit both favorable engineering selection and assessment of risk relative to alternatives and to benefits anticipated. The LMFBR base and development program will encompass a full consideration of accident situations. Finally, the construction and the operation of the LMFBR Demonstration Plant will be subject to the Commission's regulatory requirements; as required by law, a permit or license will not issue if the Commission believes such issuance would be inimical to the health and safety of the public. The Commission's regulatory review will, among other things, be based on the state of the technology at that time, and on the specific features of the design being considered. Some examples of work under way in the areas of most concern to Dr. Webb are:

a. In the area of calculational methods for determining the magnitude of disassembly accidents, Argonne National Laboratory has developed the two-dimensional VENUS reactor disassembly code. This code takes into consideration autocatalytic reactivity effects such as fuel motion. The main conclusion from this work so far is that it takes only a moderate pressure and a very small amount of material movement to cause the disassembly of a nuclear reactor. Thus during a hypothetical nuclear excursion, the minimum energy and thus the generated pressures are limited by the early occurrence of disassembly. This work has been conducted by using the FFTF parameters and characteristics. As can be seen from the referenced LMFBR Program Plan, work in this area is continuing. Because of the close coupling of potential safety problems to a particular design, a specific design (the demonstration plant for example) will be used to bring into sharp focus the LMFBR safety program, including work in the area of disassembly accidents of concern to Dr. Webb.

b. The in-pile meltdown tests performed to date in the TREAT reactor indicate that the mechanical damage potential is less than that which is thermodynamically possible by two or more orders of magnitude.

Dr. Webb uses the EBR-1 incident as a strong justification for his argument of the autocatalytic nature of fuel element melting. It has been established that the meltdown of the EBR-I fuel was due to fuel element bowing which because of the fuel's structural design caused a positive coefficient of reactivity. It is this effect that caused the short period transient in the EBR-I experiment and eventually led to the meltdown. The postmortem examination of EBR-I indicated that uranium was expelled from the core. More than half of the uranium which was originally at the core center had been pushed out by melting to a position near the edge of the core. Therefore, the EBR-I meltdown incident demonstrated that this phenomenon contributed to the shutdown of the reactor instead of leading the reactor into a "runaway" condition as asserted by Dr. Webb. In fact, the importance of fuel motion as a shutdown mechanism is also evident from recent analyses (ANL's SAS and HEDL's MELT Accident Analysis Codes) and the results from the in-pile testing in the TREAT reactor.

The following are selected excerpts from Dr. Webb's rebuttal to the AEC of July, 1973:[12]

Rebuttal

The AEC letter forwarding its Staff Review concludes that my Statement "offers no justification for reversing the AEC's current plans for designing, constructing and operating the LMFBR Demonstration Plant." However, the AEC's staff review provides no valid basis for this conclusion. Indeed, the staff review does not positively deny my allegations. . . .

I will first describe basically how the LMFBR explosion hazard arises and the main problem to be solved in predicting the explosion potential. This basic theory will hopefully enable the layman to follow this evaluation, including my original statement. . . .

The Basic Theory of LMFBR Explosion Hazard

Basically, the LMFBR contains bundles of vertical fuel rods packed together to form the "core" which produces most of the heat of the reactor. A coolant in the form of liquid metal (sodium) is pumped through the core to remove the heat and transfer it to the steam-turbine systems for electricity generation. The coolant passage space within the core is the narrow space between adjacent fuel rods. In addition, the core is pierced by non-fuel "control rods," which are used to control the nuclear reaction. Surrounding the core is a "blanket" of fertile nuclear material, again in the form of rods, which is converted to fissionable fuel (Plutonium) by the "neutron" radiation from the core. (This conversion into fissionable fuel is called "breeding.")

The explosion hazard arises because of a phenomenon called "nuclear runaway," which is an extremely rapid rise and fall in the reactor power to extreme peak levels that yields an explosive burst of energy before the "nuclear excursion" is terminated. (This is also called a "power excursion.") The reactor parameter or quantity that determines whether a runaway will be triggered is the "reactivity," and is to be controlled in order to avoid a nuclear runaway. When the reactivity is made zero, the reactor power level will remain constant; and the reactor is said to be "critical," which is the desired condition for normal, steady, full-power operations. When the reactivity is made positive (increased), but not too high, the reactor power level will rise at a controllable rate, and the reactor is said to be "supercritical." When the reactivity is decreased to below zero (made negative), the power level will decay or fall; and the reactor would be said to be "sub-critical."

But if the reactivity should increase above a threshold level, called "prompt critical," then an uncontrollable nuclear runaway will occur, which can end in core destruction, and conceivably a disastrous explosion. During the nuclear runaway the reactor is said to be "super-prompt-critical." Again, if the reactivity is below prompt critical, but still positive (above zero), the power level will rise relatively slow in a controlled rate due to the action of something called "delayed neutrons," which need not be described here. (See Appendix C for a deeper insight.) As we shall see, an unchecked *supercritical* power transient can lead to fuel over-heating and then a rise in the reactivity to trigger a super-prompt-critical power transient, or nuclear runaway.

The reactor "control rods" are the mechanical devices used to control the reactor's reactivity. They are regulated, or moved in and out of the core of the reactor (the fuel region), to control the reactivity during normal operation, in order to control and maneuver the power level. Control rod withdrawal increases the reactivity, and control rod insertion decreases the reactivity. The control rods also have a crucial emergency function to be described shortly.

The mechanisms by which the reactivity is *increased* in an LMFBR accident situation are: Fuel compaction, and perhaps something called fuel "implosion"; control rod withdrawal; and sodium-coolant expulsion or voiding from the interior of the reactor core. The mechanisms for decreasing the reactivity during an accident are: core expansion; fuel temperature rise (the Doppler Effect); and control rod insertion. Increasing, or decreasing, the reactivity is sometimes referred to as "inserting" positive, or negative, reactivity, as the case may be.

The reactivity is measured in "percent" units. About .35% reactivity is sufficient to make the reactor prompt-critical for an LMFBR (and about .7% for a water-cooled reactor). In general, a 2% reduction in the reactor core volume by fuel compaction produces about ½% positive reactivity (+ ½% reactivity). Conversely, a 2% increase in the core volume by fuel expansion produces about ½% negative reactivity (- ½% reactivity insertion). Therefore, slight compaction of the core can render the core super-prompt-critical and trigger a nuclear runaway, inasmuch as .35% reactivity equals prompt-critical. Due to the coolant space in the core, the potential for core compaction is about 50%, and therefore the potential for reactivity insertion is great; although the reactivity could not increase much beyond +1% without causing a disastrous explosion and reversal of the compaction process. Unchecked control rod withdrawal, and sodium expulsion due to sodium over-heating and boiling, can each add enough reactivity to cause a nuclear runaway, as well as fuel or core compaction.

It is the slight expansion of the core in response to the build-up of energy, and hence pressure, during a nuclear runaway that decreases the reactivity to below prompt-critical so as to terminate the nuclear runaway. (The Doppler temperature effect assists the core expansion effect in inserting negative reactivity.) Since the maximum net reactivity in a runaway will be about 1% for disastrous explosions, only the initial amounts of core expansion (about 1% increase in core volume) is needed to end even the worst nuclear runaway. If the energy generated during the runaway (called the "energy yield" or "energy released") is strong enough, the core expansion process will take the form of an explosion. The expansion of the core due to explosion will ultimately render the reactor permanently subcritical (shutdown), if we can still call a destroyed reactor a "reactor," as the core is "disassembled" by the explosion.

The severity of the nuclear runaway depends in part on the *rate* at which the reactivity increases above prompt-critical—i.e., the *reactivity insertion rate*. A higher rate means that more reactivity can be "inserted" before expansive pressures build up than the case of a lower reactivity insertion rate, which in turn means that more expansion is then required for terminating the runaway. But before the core can expand and reduce the reactivity, the fuel materials must first accelerate outward, which takes time and, thereby, delays the termination of the runaway beyond the point in time when the expansive pressures first appear. This time delay in expansion allows the runaway power level to continue to increase rapidly, and hence to increase the energy yield before expansion terminates the runaway. Since a higher reactivity insertion rate requires more expansion to stop the runaway, this time delay is lengthened, thereby worsening the energy yield. Any such delay is dangerous, since the energy yield could very quickly (of the order of a few millionths of a second) become extremely severe producing a disastrous explosion. Therefore, a greater reactivity insertion rate means more core expansion is needed to terminate the runaway, which in turn means increased time delay before a termination, which in turn means a higher energy yield and, ultimately, a greater explosion.

There is, however, another phenomenon besides the initial reactivity insertion rate on which the severity of an LMFBR nuclear runaway accident depends, and this is called an *autocatalytic reactivity effect,* which is the main focus of my concerns for the LMFBR explosion hazard, and is defined as an increase in the reactivity during or after an initial nuclear runaway due to some cause which offsets the negative reactivity inserted by core expansion and the Doppler effect. If autocatalysis occurs, the termination of the nuclear runaway will be delayed, or the runaway could even be made worse by increasing the reactivity instead of decreasing it during the runaway; or if the runaway is already terminated, a second runaway could be triggered. An autocatalytic effect, then, worsens the total energy yield in an LMFBR accident and the resultant explosion.

The LMFBR has the potential for nuclear runaway and *autocatalytic reactivity effects* because the core contains so much concentrated fuel which is *not* arranged in the most reactive configuration. This is because the fuel is arranged in bundles of fuel rods (about 0.2 inch in diameter) which are spaced apart for coolant passage. About 50% of the initial core volume is taken up by these coolant passages. The coolant passages, therefore, provide space for fuel compaction. Should the fuel over heat and melt down or slump, the core can then become compacted and insert the reactivity to trigger a runaway. Since only 2% volume reduction can raise the reactivity to prompt critical, and 2% more can result in a disastrous explosion, we can see the potential ease for runaway due to core meltdown.

Strictly speaking, any spontaneous rise in the reactivity while below prompt-critical is also "autocatalytic," as it produces a worsening power excursion, and can lead into a nuclear runaway. So, in the strict sense, any core compaction, implosion, or coolant expulsion that occur upon core overheating to increase the reactivity spontaneously are autocatalytic effects.

A core overheating and meltdown situation can be created by an "over-power accident," which I'll call a slow power excursion or rise, short of nuclear runaway, which heats the fuel at a greater rate than what the reactor coolant can remove; or by a loss-of-cooling accident in which the reactor coolant slows down as it passes normally through the core (due to loss of pumping), or is expelled from the core as it is boiled, or simply drains through a pipe rupture.

The fuel motion under meltdown can be vigorous as molten and hot solid fuel is pushed by the boiling, flashing, and exploding sodium coolant, and other high pressure forces, or as the fuel is acted on by gravity. The fuel motion upon core meltdown then determines the reactivity insertion rate at prompt-critical, which could be severe. Recall that sodium coolant expulsion due to boiling is another way which reactivity can be added to trigger a nuclear runaway. Other ways include control rod ejection and dropping a fuel rod bundle into a critical core during a refueling operation. These other ways could produce a severe reactivity insertion rate as well. (Although, it is not clear that a single control rod ejection by itself could trigger a nuclear runaway; but it could induce a power excursion, and core meltdown, and then a runaway.)

(Incidentally, the LMFBR core will contain about 250 bundles of fuel rods, all bunched together; and each bundle will contain about 200 fuel rods, making about 50,000 fuel rods total in the core. The number of control rods will be about 50, although these are much larger than a single fuel rod.)

The concern for *autocatalytic reactivity effects* arises because of the *non-uniform* nature of core meltdown and expansion. If the core were uniform and expanded uniformly as the result of a nuclear runaway, there would be no question but that the expansion would reduce the reactivity and terminate the runaway without autocatalysis. But because the expansion process will be highly non-uniform (i.e., the fuel motion will be haphazard) and because of the large amount of concentrated fuel in an LMFBR (the core contains enough fuel to make ten to forty separate "critical" reactors if fully compacted), there is the valid concern that the fuel will, on its way toward overall core expansion, collect in a different super-prompt-critical configuration long enough (of the order of 5/1000 second) to amplify the initial nuclear runaway or cause a very severe secondary runaway. These autocatalytic effects due to *fuel motion* during or right after a nuclear runaway, then, become a matter of grave concern. For an initial runaway could add enough energy to melt the whole core and even vaporize it to explosive pressures. Under these conditions, the motion of fuel can conceivably generate very severe autocatalytic reactivity effects ending in a disastrous explosion. For example, an initial runaway could be terminated by slight expansion of the core in the initial phase of the explosion. But because there is so much fuel that is relatively loosely arranged, the expansion of fuel in one region of the core could conceivably compact another region of the core and make the overall reactor super-prompt-critical again. This "explosive compaction" could make the "reactivity insertion rate" for the second runaway very high, because the reactivity is rising with explosive fuel velocities, which tends to produce an even greater runaway. Furthermore, with explosive compaction, the momentum of the fuel would be toward increasing local compaction, and, therefore, increasing reactivity, delaying the core expansion (shutdown) process until it can overcome the momentum, which would make the runaway all the more worse. The process is extremely complicated to analyze.

A special case of fuel motion is "implosion," where the fuel in the core explodes or expands inward or into an inner, hollow cavity that may have been created in the core upon meltdown. Implosion is neither compaction, nor overall core expansion; but it can be autocatalytic, as it tends to bring fuel together, like compaction, and thereby raise the reactivity. Thus implosion further complicates the calculation of core behavior in an LMFBR accident to predict whether net autocatalytic behavior is possible.

The primary purpose of evaluation of LMFBR safety, given in this rebuttal, is to convey to the layman the extreme complexity involved in calculating *fuel motion* under LMFBR accident situations or conditions, and to show that disastrous autocatalytic nuclear runaways due to fuel motion may very well be possible, and certainly have not been scientifically investigated, and that the maximum explosion potential has not therefore been established. That is, it may very well be possible for an LMFBR to suffer a disastrous nuclear explosion, releasing a large fraction, if not virtually all, of the core's Plutonium and fission product radioactivity into the Environment, as the science of LMFBRs is not well established in this regard.

So far, I have but touched on the Doppler effect, which has an important mitigating effect on the nuclear runaway. This Doppler effect *promptly* inserts negative reactivity as the fuel temperature climbs during the runaway, so as to reduce the reactivity and slow down the runaway. Without it, the explosion potential of the LMFBR would unquestionably be too high. However, the reactivity reduction potential of the Doppler effect is limited to about 1% negative reactivity, which means that autocatalytic reactivity effects conceivably could override or nullify the Doppler effect.

Another important aspect of LMFBR accidents is the "reactor scram" function, which is the rapid insertion of the reactor control rods to render the core subcritical in an emergency, and thereby avoid prompt-criticality (i.e., nuclear runaway). The SCRAM, then, shuts down the reactor so as to ensure against overheating and melting, and thus core compaction and the resultant nuclear runaway, provided that the coolant is still present to remove the "decay heat" produced by the decaying radioactivity that builds up with reactor operation. Failure to SCRAM upon detection of a core-overheat situation is expected to be the most probable way in which a nuclear runaway can occur, and the power level would remain high to effect meltdown or coolant expulsion—the main reactivity rise mechanisms.

However, once the reactor is super-prompt-critical, the control rod scram function is of no use since the runaway is extremely rapid (lasting only about 1/1000 of a second), and will be over before the control rods could be inserted appreciably. Furthermore, once the core melted-down or exploded, it seems possible that a control rod scram would not be of any help in preventing secondary nuclear runaways, as (1) the core could be so distorted as to not permit control rod insertion, since these rods are fitted into the core with little clearance; (2) the control rods themselves could be damaged or ejected by the explosion; or (3) the reactivity rise due to meltdown could override the negative reactivity "worth" of the control rod scram. In addition, there is the concern that the core could suffer overheating leading to runaway before being detected quickly enough for the SCRAM to be initiated in time to control the situation.

Finally, it is useful to compare the LMFBR with the commercial water-cooled nuclear reactor of today—the so-called "light water reactor," or the LWR. The LMFBR is greatly different than the LWR from a core meltdown and nuclear runaway standpoint. A large LMFBR has a much higher "power density in the core at normal, full-power conditions, by about 10 times (the power density is the power produced in one unit of core volume); a much greater concentration of fuel; and a much more rapid nuclear runaway given the same reactivity rise, which is a consequence of the greater fuel concentration and the different reactor coolant. LWRs have such a low fuel concentration, on the other hand, that they are not susceptible to nuclear runaway upon fuel meltdown, even if the fuel were fully compacted, according to Forbes (a point which should be confirmed). The higher power density means that the LMFBR is all the more prone to meltdown should the core suffer coolant interruption, and in that respect is more prone to nuclear runaway. The higher power density means also that the heating due to the intense radioactivity buildup in the core is greater because the radioactivity is more concentrated. This heating, called "decay heat," exists even when the reactor is subcritical, and can by itself under certain conditions cause meltdown and bring about nuclear runaway in the LMFBR. (For example, it is conceivable that the core could be distorted by an explosion such that it would not be amenable to cooling. Because of the decay heat, the core would melt down, even if the fission power level were negligible, and trigger a secondary explosion.) Nor does the LMFBR inherently shutdown (become subcritical) should the core lose its coolant, as is the case for an LWR. Instead, a reduction of coolant in the LMFBR core can by itself raise the reactivity and trigger a nuclear runaway as mentioned before; whereas the LWR requires the presence of the water coolant in the core to make the reactor critical, because of its low fuel concentration.

In other words the LMFBR has so much fissionable material in concentrated form that it is prone to suffer nuclear runaway and explosion accidents if the core configuration or condition is perturbed slightly. Indeed, a mild local perturbation in the core of an LMFBR could generate a strong enough over-power transient so as to melt down the entire core and lead to an even stronger nuclear runaway, the bounds of which have not been scientifically determined. Again my concern for *autocatalytic reactivity* effects is that a core undergoing a nuclear runaway may possibly be capable, during an early phase of the explosion, of either compacting or imploding part of its fuel so as to amplify the initial nuclear runaway or to trigger stronger secondary nuclear runaways that end in a disastrous explosion. Core explosion is given the name "core disassembly," although this term could imply relatively nonviolent core disruption or expansion as well. Core disassembly is the reverse of compaction or implosion and eventually stops the nuclear runaway by virtue of the fact that the fuel is blown apart so that it can no longer sustain an atomic fission chain reaction to generate energy. But, if the energy created by the runaway is great enough, the disassembly would occur explosively. It is crucial to predict the *fuel motion* during the accident to determine whether the fuel will implode or compact in an autocatalytic manner, or whether the fuel disassembles permanently without chance for re-assembly into a critical mass, and runaway, later on.

Complicating a prediction of the motion of fuel, and thus the strength of nuclear explosions (I shall use the term "nuclear explosion" to denote the combination of the nuclear runaway and the explosion which follows.), is the existence of a myriad of different pressure sources, such as sodium coolant boiling, which can itself be explosive, gaseous by-products of the fission process, and fuel vapor and other effects, all of which are inter-related and dependent on the conditions of the reactor at the onset of trouble. These complications, plus the difficulty in predicting theoretically whether autocatalytic reactivity effects due to the complicated fuel motion can occur, and then confirming the theoretical predictions experimentally, is the central problem which my Statement, and this Rebuttal of the AEC's comments, address.

Finally, we present Appendix C of Dr. Webb's rebuttal to the AEC. Webb suggested (in his rebuttal remarks above) it be referred to "for a deeper insight." It appears that Appendix C would still be lying in a drawer at the NRC had we not searched for the record of his submittal remarks to the AEC. In our opinion, Webb's Appendix C may well be the most concise, accurate, and sober description of the potential explosion risks associated with fast-breeder reactors that is suitable for consideration by the public today.

Appendix C of Webb's Rebuttal

Basic Theory of LMFBR Nuclear Runaway in More Detail

A nuclear power reactor, such as an LMFBR, generates energy or heat for eventual electric power production by the fissioning (splitting) of uranium and plutonium fuel atoms. This fissioning is caused by the interaction of fuel atoms with small atomic particles, "neutrons," which fly around inside the reactor at great speeds. When a neutron strikes the nucleus of a fuel atom, it is likely to be absorbed and cause the atom to fission. The number of fuel atoms in the core is extremely large; and only a tiny fraction of these are fissioned in one second. Numerically, one ton of fuel in a large 1000 MW LMFBR is made up of about 2×10^{27} atoms, i.e., 2 thousand trillion trillion atoms. In one second our 1000 MW LMFBR will fission 3×10^{19} fuel atoms, or 3 billion trillion atoms. Hence to fission all of the fuel atoms in a ton of fuel in our 1000 MW LMFBR would require $2 \times 10^{27} \div 3 \times 10^{19} = 2/3 \times 10^{8}$ seconds (67 million seconds), or about 3 years. Therefore, when I speak of fissioning, extremely large numbers are involved, even though I might refer to one or a thousand fissions. Likewise, large numbers of neutrons are involved.

Each fission, besides releasing the sought after energy, releases several neutrons (2.5 neutrons per fission on the average), which are then available to carry on the process through the next fission cycle in order to sustain the fissioning rate (power level) in the reactor. However, since only one of the released neutrons is needed for the "next fission," 1.5 neutrons per fission are extra, the difference between 2.5 and 1.0. But as we shall see next, these extra neutrons are lost to the system either by leakage or non-fissionable absorption, except for slight imbalances which give rise to power level transients, which can be slow or extremely rapid, as in an explosive nuclear runaway.

Because of the finite size of the batch of fuel in the reactor, which is called the "reactor core," a fraction of the neutrons produced by fission are lost due to leakage—i.e., some neutrons escape the core and never return to cause fissions. Because, too, non-fuel materials exist in the core which absorb neutrons, such as structural materials and Uranium-238, used to dilute the fuel, some of the neutrons are absorbed without causing fission. The result of the size and non-fuel effects is a competition between losses (leakage and absorption) and gains (fission neutrons). When these competing factors are balanced, the fission rate or power level is constant, and the reactor is said to be "critical." In general, whenever fissionable material in a critical reactor is brought closer together (fuel compaction), the chances for the neutrons striking fuel atoms and causing fission, rather than leaking out of the core, will improve; and the neutron balance in the fission cycle tips in favor of excess neutrons available for fissioning. The extra neutrons then produce fissions which in turn produce extra neutrons, and so on as the fission-neutron cycle repeats. The result is a growing neutron population and a growing fissioning rate, and hence an increasing reactor power level. In this condition the reactor is said to be "supercritical." In the reverse case, when fuel expands (fuel moving apart), the neutron leakage increases; and then the neutron balance tips the other way, causing the power level to decay, since less than one neutron released per fission on the average is available to sustain the next fission. In this condition the reactor is said to be "subcritical."

The percentage difference between the number of neutrons available for fissioning and the number needed to sustain the fissioning rate at a constant level is a crucial parameter called the "reactivity." Therefore, when the reactivity is positive, the fissioning rate grows and the reactor is supercritical; and when the reactivity is negative, the fissioning rate decays, and the reactor is subcritical. Thus, *fuel compaction increases the reactivity, and fuel expansion decreases the reactivity.* When the reactivity is zero, the reactor is critical. As we shall see, +1% reactivity is very strong.

There is another kind of neutron balance involving the *time scale,* and concerns the *controllability* of reactor power level increases. Foremost is the "neutron lifetime," which is the time period between the release of neutrons from one set of fissions until these fission neutrons cause the next set of fissions (a fission cycle). The neutron lifetime is extremely short in an LMFBR—about .0000001 seconds, or one-tenth of a millionth of a second, due mainly to the *fast* speeds of the neutrons, which is why the LMFBR, Liquid Metal Fast Breeder Reactor, is called a "fast" reactor—meaning a fast neutron reactor. If this is all there were to fission-physics, then once a reactor was made slightly supercritical, it would quickly runaway with an uncontrollable burst of energy. In order to appreciate this, assume that a large 1000 megawatt LMFBR was critical at a feeble power level of 1/100 watt, which would be .00000000001 of the reactor's designed full-power level. Then assume the reactor is made supercritical by a slight compaction of the fuel so that the reactivity is increased to +.5%. (Roughly, a 2% reduction in reactor core volume by core compaction adds .5% of positive reactivity. Potentially, the core volume could be reduced by about 50% by compaction; but as we shall see, a nuclear runaway would explode the core before it could be compacted very much past a 2% volume reduction). A reactivity of .5% means that the number of fissions per cycle would increase by .5% with the passage of each neutron lifetime (i.e., from one fission cycle to the next fission cycle). This means that the number of fissions occurring per cycle increases, not at a steady rate, but at progressively increasing rate (i.e., "exponentially"). This is because the number of fissions in one cycle is .5% greater than the number of fissions in the immediately preceding fission cycle, and not .5% of the number of fissions in the first cycle after the reactivity was raised above zero. That is, the *increase* between successive fission cycles is .005 times the *current* number of fissions occurring per cycle. Since the *increase per cycle* gets larger when the *current number of fissions per cycle* gets larger, the *growth rate* of fissioning accelerates, instead of staying constant, as time progresses.

As an illustration, let us assume that the cycle produced 1000 fissions, and then compare the case of steady-rise with the exponential-rise after 10, 100, 300, 1000, and 2000 cycles, respectively, given the .5% reactivity. The following table illustrates the difference between the two cases.

Number of Fissions Occurring in the "Nth" Cycle

Nth Cycle	Steady Rise	Exponential Rise
1st	1,000	1,000
10th	1,050	1,051
100th	1,500	1,650
300th	2,500	4,500
1,000th	6,000	143,000
2,000th	11,000	20,500,000

From the table we see that there is little difference in the first 100 cycles. However, the number of fissions per cycle in the exponential case begins to get progressively greater than the steady-rise case, until past the 1000th cycle when the exponential rise "runs away." This process happens extremely quick in time because of the short neutron lifetime (time period of the fission cycle). For example, there are 50,000 fission cycles in just one-thousandth of a second, or millisecond, which allows a tremendous growth in fissioning in a very short interval of time.

Let us now ask what would be the power level and energy generated after our hypothetical reactor was supercritical at .5% reactivity for one millisecond. The answer is that the power level would grow, if it were not controlled by core expansion (and fuel burn-up) to 500 billion times the 1000 megawatt full-power level designed for the reactor, starting with only a feeble 1/100 of a watt; and the energy generated during the millisecond would be 100 billion megawatt-seconds, roughly equivalent to a 25 megaton nuclear weapon explosion. *Actually*, the heat generated early during the transient would create pressures within the fuel to expand the fuel, which decreases the reactivity to a negative value. (Just as the fission rate grows exponentially when the reactivity is positive, the fission rate decays exponentially when the reactivity is negative. Therefore, when the reactivity is negative, the power level will quickly decay to a feeble level with the same rapidity as the runaway rise in power level.) This expansion, therefore, affects the reactivity, and the course of the runaway, and must be taken into account. When it is, an LMFBR under a .5% reactivity runaway (and no Doppler feedback) will produce an explosion of the order of 1000 lbs. TNT equivalent, excluding autocatalytic reactivity effects, according to estimates. This negative reactivity effect due to expansion thus terminates the runaway, limiting it to a much less violent explosion—about 1000 lbs. TNT equivalent for the assumed reactivity condition in an LMFBR. This phenomenon of exponential growth of the fission rate is called a "nuclear runaway," which can produce a burst of explosive energy.

Given such hypothetical reactor behavior, the reactor would not be controllable, since a slight increase in the reactivity, which a reactor operator would normally want to make in order to raise the power level from shutdown to full power level, for example, would lead instantly (within a millisecond) to reactor destruction before the control equipment could respond. This is because the mechanical reactor control equipment couldn't make the super-fine changes in reactivity that would be needed to raise the reactor power level at a controlled rate for our hypothetical reactor. That is, the nuclear runaway would be over within a millisecond, before the control rods would move any appreciable amount.

Fortunately, for *control* purposes, a small fraction of the fission-released neutrons (about .3% to .7%) in a *real* reactor do not appear promptly with the fissions, but are emitted by the fission fragments with about a one second *delay*. The fraction of the fission neutrons which are delayed is called the "delayed neutron fraction." If a reactor was made supercritical, *but with the reactivity kept below the delayed neutron fraction*, the delayed neutrons would have the effect of suppressing the growth rate of the fissioning, enabling one to control the reactor. To understand why, consider again our hypothetical supercritical reactor *with no delayed neutrons.*

With the reactivity positive, there would be more fission-released neutrons to cause further fissioning than would be needed to sustain the fission rate at a constant level. But by not being delayed, the extra neutrons would cause the extra fissioning within the short neutron lifetime. Hence, the fission rate would rise extremely rapidly in an exponential, runaway fashion. But if the *extra* neutrons were delayed by about one second, then the extra fissioning, caused by these extra neutrons, would be correspondingly delayed. The result is that the fissioning rate, or reactor power level, in a real reactor would grow slowly, over the time scale of seconds instead of 1/10 of a millionth of a second (i.e., instead of in the runaway fashion, if the reactivity is less than the delayed neutron fraction). In this state the reactor is *still* said to be "supercritical." This neutron delay, then, provides enough time for the reactor control system to maneuver the power level during normal operation. When the desired power level is reached, the reactivity is returned to zero, so that the reactor will be made critical—i.e., producing power at a constant level.

However, if the reactivity is raised to exceed the delayed neutron fraction, then there will be an excess of prompt neutrons available for extra fissioning. The growth of fissioning will then occur over the short time period of the "neutron lifetime," instead of over a long delayed period.

Hence, when the reactivity exceeds the delayed neutron fraction (about .35% in an LMFBR), a nuclear runaway will ensue in the fashion of our "hypothetical" reactor previously discussed. In this runaway condition, the reactor is then said to be "super-prompt-critical." When the reactivity equals the delayed neutron fraction, the reactor is said to be "prompt critical," which is the threshold for nuclear runaway. *The crux of reactor control is to keep the reactivity below prompt critical, or else an explosive nuclear runaway will occur.* But this is not always possible, as an accident could make the reactor super-prompt-critical.

Next, we shall summarize the phenomena which can change the reactivity, as these reactivity effects are crucial to the control and the accident behavior of the LMFBR. These phenomena are as follows:

- *Reducing the neutron leakage increases the reactivity.*
  This is accomplished by bringing fuel together (compacting fuel or adding more fuel) so that the neutrons have a better chance of interaction with the fuel atoms, rather than being lost due to leakage. A special case of compaction is implosion; e.g., when the fuel explodes into a hollow, interior cavity, while being essentially confined from exploding outward. A fuel meltdown could produce core compaction.
- *Increasing the neutron leakage decreases reactivity.*
  This is accomplished by moving fuel apart: expansion as with explosion; fuel falling away from the core; or fuel from the core being removed mechanically or carried away by the flowing coolant.
- *Increasing the neutron absorption by non-fuel material decreases reactivity; conversely, reducing such absorption increases reactivity.*
  This phenomenon is used to control the reactor once enough fuel is assembled to make the reactor critical. The control is effected by inserting or withdrawing "control rods" into and out of the reactor core. These control rods are made of non-fuel, neutron-absorbing material. Thus inserting them into the core robs neutrons that would otherwise cause fission, and thereby, decreases the reactivity. Withdrawing the control rods reduces the non-fuel absorption of neutrons and increases the neutrons available for fissioning, and thus increases the reactivity. In general, the reactor is designed so that the neutron

balance is achieved when the control rods are withdrawn to the "critical height" position that is part way out of the core. When the control rods are withdrawn to this height, the reactor will be critical. Further withdrawal will make the reactivity positive, and the reactor will be supercritical. If the control rods are withdrawn too far, the reactivity can increase beyond the delayed neutron fraction, and the reactor will be made super-prompt-critical, and then a nuclear runaway will ensue. These control rods are regulated so as to raise and lower the reactor power level for normal operation while keeping the reactivity below prompt critical. Also, as the fuel "burns-up" with use (each fission destroys a fuel atom), the reactivity would tend to become negative (i.e., make the reactor subcritical) since fuel burn-up has the effect of removing fuel. (A subcritical reactor could not produce power because the power level would decay to practically zero.) To compensate for this burn-up effect, the control rods are withdrawn slowly over the period of months as the fuel is depleted to keep the reactor critical and producing power. The fuel will continue to be depleted with reactor operation until the control rods are fully withdrawn from the core, in which case the reactor power level could not be sustained for normal operations (end of life), and the reactor would have to be "refueled." However, if the reactor suffered fuel meltdown in the "end-of-life" condition, there is still the reactivity rise potential due to core compaction and, therefore, the potential for nuclear runaway accidents. The control rods also have a crucial safety function. In the event that the reactor should reach a dangerous reactivity condition (near prompt critical) the "protection system" is designed to rapidly insert or "scram" the control rods to render the reactor subcritical. This safety action is called "reactor scram."

- *Increasing the fuel temperature decreases the reactivity.*
  This is an inherent safety mechanism called "Doppler feedback," which is being designed into LMFBR's in the United States. It is designed to act *during a nuclear runaway* to limit the energy burst, when a control rod scram would be too slow to have any mitigating effect. More specifically, as the temperature rapidly increases in the fuel during a nuclear runaway, the Doppler effect promptly subtracts reactivity to slow the runaway and, in some mild runaway cases, can render the reactor safely subcritical until the control rod scram can permanently

shut down the reactor without the generation of excessive temperatures (i.e., explosive pressures). However, in most runaway accidents, the source of the initial reactivity increase which caused the runaway will persist to override the negative Doppler reactivity. Other sources of positive reactivity may occur as well. So Doppler feedback is not sufficient to stop most accidents. Also, the Doppler reactivity reduction potential is limited practically to about 1% of negative reactivity. Thus Doppler is not enough to cope with the potential for accidental positive reactivity addition. (The negative reactivity of overall core expansion is being counted on as the main shutdown mechanism for terminating a nuclear runaway.) The chief role of the Doppler, then, is to slow down the nuclear runaway long enough to enable the expansion process and make subcritical. This mitigating effect of the Doppler can be strong.

• *Sodium coolant (liquid metal) expulsion from the core can increase or decrease reactivity, depending on which regions of the core are made devoid of coolant.*
This effect is due to a trade-off between increased neutron leakage and increased neutron absorption by the fuel when coolant is "voided" from the core. The net reactivity change can be positive if the sodium coolant is expelled (voided) from the inner regions of the core, where neutron leakage from the core is lowest.

Having now described the basic reactivity change mechanisms, let us learn how these mechanisms can be called into play in an LMFBR accident to bring about a nuclear runaway and explosion.

The fuel in the LMFBR is arranged in bundles of fuel rods spaced somewhat apart for coolant passage (heat removal). Therefore the fuel is not arranged in its most reactive state, since the coolant passages provide space for fuel compaction. However, the reactor fuel rods are designed to be fairly rigid so that they won't bow inward or slump (compact) during normal operations and add excessive reactivity. However, if the fuel should over-heat, either by unchecked control rod withdrawal, which adds reactivity and causes the power level to rise to excessive levels, or by a loss-of-coolant, the fuel will melt, lose its rigidity, and could then collapse onto itself as the molten fuel moves into the coolant passage space. The result of core meltdown, then, could be core compaction, which can cause an excessive rise in reactivity. Keep in mind that it takes only slight compaction to raise the reactivity to prompt critical—about 2% volume reduction of the core; and then slightly more compaction to trigger the nuclear runaway. That is, slight fuel movement either way can have either a serious positive reactivity effect, or a strong negative reactivity, shutdown effect. Actually, after the reactor has operated a while, intense radioactivity builds up, so that even if the reactor was made subcritical and the fission power level dropped to feeble levels, the heat from the decaying radioactivity called "decay heat," which is substantial, will persist. This decay heating can by itself melt the fuel and could bring about core compaction.

Besides fuel meltdown, sodium coolant voiding can trigger a nuclear runaway as well. For example, a loss-of-coolant flow accident or over-power accident can lead to coolant overheating, boiling, and then expulsion or voiding of the coolant from the core. This sodium voiding can then add reactivity past prompt critical to produce a nuclear runaway. This is an example of autocatalytic behavior, where an LMFBR accident feeds itself a dose of positive reactivity by overheating to produce a nuclear runaway, which then worsens the accident.

The central concern in LMFBR accident analyses is the behavior of the *reactivity* during the accident. From the foregoing it is clear that besides *coolant voiding* we must be able to accurately predict *fuel motion* during an LMFBR accident situation to determine whether the explosion process itself can compact part of the core to a sufficient degree to *increase* the reactivity before overall core expansion permanently renders the reactor subcritical or shutdown. If sufficient fuel compaction occurs during an explosion to offset the negative reactivity due to Doppler and overall core expansion, then the net reactivity can increase, instead of decrease during the nuclear runaway, and the runaway will become worse (faster), instead of being terminated; or if the nuclear runaway had already been terminated, a second one could occur. As we've seen at the outset, the energy can build up very quickly to dangerous, explosive levels when the nuclear runaway condition is *prolonged*. The behavior of the reactor when reactivity rises instead of falls during the accident is called "autocatalytic," meaning that the core is its own catalyst—speeding up its own fission reaction rate. Conceivably, *autocatalytic reactivity effects* could even exhaust the Doppler negative reactivity effect, which would make an explosion all the more severe. Eventually, however, overall core expansion (explosion) would take over and drive the core subcritical. The question is, though, how much energy can the nuclear runaway(s) generate before being finally terminated—the energy being then correlated with the size of the resultant explosion.

The energy yield of an LMFBR nuclear runaway accident, which is the measure of the force of the explosion, is related to the rate at which the reactivity rises above prompt critical, i.e., the "reactivity insertion rate." If the rate is low, the nuclear runaway will proceed less rapidly than otherwise, giving the fuel material time to accelerate outward (expand) and provide the offsetting negative reactivity before too much reactivity builds up to generate a stronger runaway. If the rate of reactivity increase is high, then more reactivity can be "inserted" before the expansion occurs, and a stronger runaway occurs. Remember, it takes time for fuel material to accelerate and expand, which allows for reactivity insertion. Initial meltdown events are characterized by upper limits of reactivity insertion rates of about 200% per second, which when mitigated by the Doppler effect, yields the 500 lb. TNT-order explosion, assuming no autocatalysis. But autocatalytic reactivity effects such as explosive compaction could conceivably yield insertion rates in excess of 1000% per second. Therefore, fuel motion is the primary object of study in LMFBR analyses, and must be fully understood to establish the maximum explosion potential of the LMFBR.

Complicating the nuclear runaway problem is the amount of fuel concentrated in an LMFBR, which is enough to make somewhere between 10 to 40 separate critical reactors, if the fuel is fully compacted (fully dense). Thus for example a nuclear runaway could be terminated by slight expansion of core materials during the initial phase of a nuclear runaway explosion, only to compact enough fuel later on to return the core, or a part of it, to super-prompt-critical; i.e., to trigger secondary nuclear runaways. However, with explosive compaction, the rate at which the reactivity would increase would be great, and the momentum of the compacting fuel would have to be overcome, which delays the *shutdown* reactivity and conceivably could enable the runaway to grow to very dangerous levels.

These factors, then, make explosive compaction a matter of grave concern. (Indeed, the atomic bomb is produced by explosive compaction [the compaction is affected by detonating a TNT charge].) Whether an LMFBR can be made to explode like an atomic bomb is a question I honestly don't know the answer for. All I can say is that I have seen no analyses which rule out the possibility; and that I'm prevented from learning the physics of the atomic bomb, since the information is kept secret. My best judgment, though, is that the worst autocatalytic nuclear runaway in an LMFBR would not produce an atomic-bomb-like explosion, but that it may produce a severe enough explosion to "blow-up" the reactor and allow the escape of the radio-activity to the environment (the worst conceivable LMFBR explosions mentioned in this rebuttal range from 500 lb. TNT equivalent to the order of 20,000 lb. TNT, which compares with a 20,000 *tons* of TNT equivalent for the first A-bomb).

It is useful to compare the commercial, water-cooled reactors now being operated—the so-called light water reactors (LWRs)—with the LMFBR. The concentration of fissionable fuel in an LMFBR core is much greater than the LWR. In fact, the LWR fuel concentration is so low that without the water coolant, the fuel probably cannot be made critical even if the fuel is fully compacted. It turns out that the LWR fuel can only be made critical if the fuel is spaced apart in the form of fuel rods with water in between. Unlike the sodium coolant in an LMFBR, the water in an LWR greatly slows down the neutrons, which are released at high speeds by the fissioning. A slow neutron has a much better chance for splitting atoms than a fast neutron. Hence, a lesser fuel concentration is needed in an LWR. But if the water coolant should be expelled or drained from the core of an LWR, the reactor would be rendered subcritical, since the fission neutrons could not be slowed down, and without the slow neutrons the low fuel concentration could not sustain the fissioning. In contrast, the loss-of-coolant accident in an LWR presents the danger of a core meltdown, and the associated possible disaster of the built-up radioactivity escaping to the environment (due to the meltdown causing a breach in the reactor container). But because the LWR has a low fuel concentration, it does not have nearly the reactivity or nuclear runaway problem associated with fuel meltdown or coolant expulsion in an LMFBR.

Further, the Doppler effect is stronger in the LWR, and the neutron lifetime is longer by a factor 1000. These facts make a nuclear runaway in an LWR less severe compared to an LMFBR for the same initial reactivity condition. (However, the LWR still has a serious potential for nuclear runaway; but this fact is beyond the scope of this LMFBR safety review.) Finally, the LMFBR has a power density in the core that is about ten times higher than that of an LWR. The power density is the amount of heat (power) generated in a given volume of core. This higher power density means that core meltdown occurs more vigorously, should adequate cooling be lost, than in an LWR. Also, the "decay heat" in an LMFBR is correspondingly stronger, which makes core meltdown worse than in an LWR without adequate cooling. This decay heat is troublesome for a number of reasons, one of which is that even if the LMFBR had shutdown (subcritical) after suffering a meltdown, the fuel might freeze into an uncoolable mass, which could soon melt again, generating the possibility of re-assembly back into a "critical mass" and nuclear runaway.

Notes

1. Microfilm or full size copy of PhD thesis is obtainable from University Microfilm, Ann Arbor, Michigan. See *Dissertation Abstracts,* Vol. 33, No. 2 (Aug. 1972), pp. 754. B–755.

2. LMFBR Demonstration Plant, Environmental Statement, WASH-1509, United States Atomic Energy, April 1972. p. 118.

3. N. Hirakawa and A. E. Klickman, "An Analysis of the KIWI-TNT Experiment with MARS Code," *Journal of Nuclear Science and Technology,* Volume 7, No. 2, pp. 1–6, January 1970.

4. R. E. Webb, "Critical Review of the KIWI-TNT Power Excursion as Calculated by the MARS Fast Reactor Excursion Code," Draft paper, unpublished.

5. V. Z. Jankus, "Calculation of the Energy Yield in the KIWI Transient Nuclear Test, KIWI-TNT." ANL-7310, p. 366.

6. SEFOR Reports: GEAP-13598, 10010-24, 29, 30 and 31. See Refs. 2 and 10 for discussion of SEFOR in regards to LMFBR safety. Ref. 2, pp. 19–20.

7. R. E. Shaver and N. G. Wittenbrock, "Review of Reactor Safety Analyses of Fast and Liquid Metal Cooled Reactors," BNWL-477, UC-80, Reactor Technology, November 1967, pp. 21, 35.

8. T. J. Thompson and J. G. Beckerley, *The Technology of Nuclear Reactor Safety,* Volume 1, MIT Press, p. 616.

9. J. B. Nims, "Fast Reactor Meltdown Experiments," *Proceedings of the Conference on Breeder, Economics and Safety in Large Fast Power Reactors,* October, 1963, ANL-6792 (December 1963), pp. 203–231.

10. LMFBR Program Plan, WASH-1110, Volume 10, Safety, pp. 10-213–10-23.5.

11. WASH-1101-1110. LMFBR Program Plan, August 1968. (It is presently being updated.)

12. Richard E. Webb, *The Explosion Hazard of the Liquid Metal Cooled, Fast Breeder Reactor (LMFBR): And the Unconstitutionality of the AEC's Civilian Nuclear Power Program,* July 1973 https://www.nrc.gov/docs/ML1508/ML15083A202.pdf. The authors would like to thank copyeditor Gary Morris for locating the digital record of this document.

CHAPTER 6

Tickling the SEFOR Dragon

Between the time at which positive reactivity is inserted and control rods begin to move, the reactor behavior depends primarily on the Doppler effect to limit the severity of the accident.

The integrated negative Doppler reactivity from operating power level to the point of fuel rupture is required to be large compared to credible outside sources of rapid reactivity insertion.

. . . Accidents in which the control system does not function are considered hypothetical. In such accidents, the core will be destroyed, and the question of primary interest is the containability of the accompanying energy release.

K. P. Cohen, General Electric Company

It is, in our view, unlikely that one will be able to design for the worst accident permitted by the laws of nature and end up with an economically interesting system, even after extensive additional research and development has been carried out.

P. M. Murphy, General Electric Company

As K. P. Cohen and P. M. Murphy were two of General Electric Company's principal officers in charge of the LMFBR program, it appears that GE had no illusions about the risks that the SEFOR project involved. SEFOR was an experimental reactor; it was never intended to suggest a reactor design that would be an "economically interesting system." The scientists and engineers at GE appear to have designed it with three principal performance features in mind:

- As there were a number of ways that positive (or negative) reactivity could be added to the reactor core by intention or accidentally, SEFOR was designed to minimize the potential for all known positive and negative reactivity additions save two: a planned intentional positive reactivity addition that could be accurately controlled in magnitude and limited to a specified time period, and the theory-predicted Doppler effect negative reactivity provided by the plutonium-oxide fuel. These design features were required to eliminate, as carefully as possible, all of the effects of reactivity changes in the core, save these two, in order to allow an accurate measurement of the Doppler effect.
- The reactor was designed so that a highly controlled and accurately specified amount of positive reactivity could be "inserted" into the fuel core. This task was performed with the Fast Reactivity Excursion Device (FRED).[1]
- The design was intended to minimize the possibility that the positive reactivity *intentionally* added to the reactor could exceed the amount of *inherent negative reactivity* that was predicted to occur due to the Doppler effect. This feature was critical; the design sought to ensure that the positive reactivity added could not *override* the negative reactivity provided by the Doppler effect.

The primary goal of the SEFOR program was to quantitatively demonstrate the Doppler effect without suffering a damaging explosion. A principal design goal was to ensure that the fuel temperature increase that would result from the planned reactivity addition would not result in any fuel melting.

Reactivity Effects Important in the SEFOR Experiments

Reactivity: The excess (positive or negative) in the number of neutrons produced per neutron lost during an average lifetime in a fast reactor—about one 10-millionth of a second.[2]

- The primary fission reaction of plutonium results in "splitting" of the 239 isotope into two parts with the production (typically) of either two or three neutrons.

- The fission of one atom generates 207.1 million electron-volts, or 0.00000000003318 joules of energy.
- 1 gram of plutonium contains about 2.52 billion trillion atoms.
- 1 gram of plutonium fissioned produces, if neutron production per fission is assumed to be exactly two, 2.52 billion trillion (new) neutrons—accompanied by an energy release of approximately 20 tons (40,000 pounds) of TNT. It follows that 1 kilogram of plutonium fissioned releases energy equivalent to approximately 20,000 tons of TNT (Nagasaki bomb yield).
- During normal reactor operation, a slight excess of neutrons is produced by fission above the combined number of neutrons that are absorbed by the fuel or lost through the surface of the fuel. This "reactivity" determines the chain-fission-reaction rate in the fuel.
- Depending on the type of reactor, the excess neutrons can be of two general types: slow and fast; the former having velocities (on average) of approximately 2 km/s (~4,500 miles per hour) and the latter having velocities (on average) thousands of times higher.
- The reactor is normally controlled by adjusting (inserting or removing) neutron-absorbing "rods" that effect a reduction or increase as necessary in the reactivity to maintain the desired operating power level (number of fissions per unit time).
- So-called "transient" increases in reactivity, which can be caused by normal variations in the chain reaction rate as well as by accidental occurrences, are normally automatically controlled by positioning slow-acting control rods in the reactor fuel or, as in SEFOR, positioning a reflector "curtain" surrounding the reactor core.
- If accidental increases in reactivity occur beyond these "normal" increases, a separate, automatic, fast-acting SCRAM control system is provided to shut the reactor down before the core is damaged by overheating.
- If the number of fast neutrons in the core become sufficient to maintain the reactor fuel in a supercritical reactive condition (fissioning by fast neutrons only), such "prompt" neutron chain reaction rates can increase by a factor of several thousand, producing nuclear fission bomb explosion intensities.
- In the event that such a "super-prompt-critical" condition is reached, the only resort is to SCRAM as quickly as possible.

- Enter the Doppler effect. The Doppler effect, a prompt-negative-reactivity-effect, was designed into the SEFOR reactor by using (as fuel) the oxide of plutonium, rather than pure metal plutonium as normally required for nuclear weapons explosion yields. The Doppler prompt-negative-reactivity addition occurs as the temperature of the fuel increases; if sufficiently strong, the Doppler effect will slow (cancel out) a positive super-prompt-critical reactivity addition, providing critical time for movement of the SCRAM rods.

Quoting GE's Dr. Cohen: "The integrated negative Doppler reactivity from operating power level to the point of fuel rupture is required to be large compared to credible outside sources of rapid reactivity insertion," and, "Accidents in which the control system does not function are considered hypothetical. In such accidents, the core will be destroyed, and the question of primary interest is the containability of the accompanying energy release."

In lay terms, Cohen's statements mean that the amount of explosive "reactivity" introduced into the reactor core must, at a minimum, be sufficiently compensated for by the amount of reactivity removed by the Doppler effect in order to prevent the fuel temperature reaching its melting point. The fuel melting point limit is to be avoided to obviate the possibility that melting could alter its spatial arrangement in such a way as to cause the reactivity to increase as a result of *autocatalytic* effects discussed by Dr. Webb in chapter 5. Finally, Cohen is stating that if the Doppler effect is not sufficiently strong to effectively cancel the positive reactivity addition, the only result is to activate the emergency SCRAM system. If the emergency SCRAM does not function in these circumstances, the core will be destroyed, and the question of primary interest is whether the containment provided for the reactor can withstand the energy released to prevent failure of the containment—resulting in a worst-case accident that could cause massive releases of radioactivity to the environment.

The What If Question

We repeat the statement from chapter 2 that considered SEFOR Prompt Critical Transient Experiment No. 6 conducted during the last week of December 1971.

If, in Test No. 6, the increase in SEFOR's power level caused by the positive reactivity insertion had not hesitated (as predicted by the Doppler effect); if the reactivity insertion had been accidentally maintained longer than the planned 0.1-second duration; and if the SCRAM procedure had failed—there was the real possibility that the power level might have increased to levels with the potential to rupture the containment.

We note that Test No. 6 left two critical questions unanswered:

- The experiments did not address the potential hazard of nuclear explosions that might be possible if sufficiently large amounts of prompt-critical reactivity were "inserted" into the core by accident or as a result of natural disaster that could override the Doppler effect. *Dr. Webb's principal concern, the possibility of unpredictable autocatalytic reactivity additions, was not addressed.*
- The possibility of such autocatalytic effects occurring made all the more critical the reliability of the SCRAM mechanisms provided for the reactor. It appears that the reliability of the SCRAM mechanism was effectively *assumed*.

During the last year given to drafting this book, while continuing research into the general subject of nuclear safety regulation and fast nuclear reactor safety science, we found the following statement written by Dr. David Okrent, now deceased. Dr. Okrent was a leading authority on nuclear safety research and regulation and long-term influential member of the Advisory Committee for Reactor Safeguards:

> The Staff report references various experts who have estimated an unreliability of SCRAM from 10^{-3} to 10^{-4} per demand. At the ACRS Subcommittee meeting on August 26, 1970, General Electric stated that experience with GE reactors led to a failure probability of 8×10^{-4} with a 95% probability. It was stated that to demonstrate empirically an unreliability of 10^{-7}, approximately 300,000 reactor years with a zero failure history would be required. . . . In fact, during the recent past, another failure has actually been experienced at Hanford and a partial failure at SEFOR thereby reinforcing the Staff position.[3]

The SEFOR SCRAM System

SEFOR was controlled by movable radial reflector segments that surrounded the reactor core outside of the reactor vessel. The reflector was divided into ten segments that could be raised or lowered vertically to give the desired control. Reflector control was feasible in SEFOR because the core was relatively small with a large radial neutron leakage component (about 21% of all neutrons). This type of control was desirable for SEFOR for several reasons, the primary one probably being maximum safety during the super-prompt-critical experiments that were a part of the planned experimental program— placing the primary reactor control and shutdown system in a less vulnerable location outside of the reactor vessel provided an added margin of safety.

The 6-inch-thick, 34-inch-high nickel alloy cylindrical reflector was divided by radial cuts into equal sectors. Each sector comprises the active portion which was raised and positioned by its drive between the core and the poison blanket that surrounded it.

The reactor was scrammed by simultaneously dropping the reflector segments to a position that placed them below the bottom edge of the core.

The reflector guide extended from the top of the core down to the ceiling of the drive cell. It was composed of two concentric shells joined by radial webs that supported the segment guide rails and formed the channel within which the reflector segments were moved.

Two types of drive mechanisms were used to position and scram-control the reflector segments, two for fine and the other eight for coarse control. The coarse control drive normally fixed its control rod at either end of its stroke, although the rod segment could be interrupted at any intermediate point. The fine drive was continuously and accurately positionable throughout its stroke.

The SEFOR Partial SCRAM Event

Quoting from the SEFOR Sixth Quarterly Plant Operation Report:[4]

> Reactor operation at Five MW for Test Procedure Group III, Static Tests was in progress. *Tests with main primary and main*

secondary flow rates of 800 gpm had been completed and the main secondary flow rate was being increased by movement of the flow controller setpoint when a reactor scram occurred at 1355 on September 12, 1970. (emphasis added) Main secondary flow at the time of scram was approximately 1400 gpm. Flow fluctuations were about +/-25 gpm. The reflectors dropped approximately 5 cm, carriage separation occurred on the fine drives, power dropped to approximately 3 MW, when an automatic scram reset occurred. An annunciator alarm and the scram event recorder indicated "Low Flow Main Secondary." No other event was recorded or observed before or during the scram.

The operator immediately pushed the manual scram button, and the scram was completed. A low pressure freon header trip had been inserted previously in the safety system since one freon unit was not required for the 5 MW operation. The short duration trip from the low flow-main secondary completed the two-out-of-three logic for scram. Subsequent investigation revealed that the main contacts on the K1 (scram solenoid) contactor were opening a noticeable time before the "Hold-in" (or auxiliary) contacts (through which the contactor coil current flows). With this relative opening of contacts on the contactor, if a trip signal consisting of a short duration pulse were received by the scram relay (mercury wetted contacts with time to open of 3 to 4 milliseconds), the voltage could be removed from the scram bus, the main contacts could open, the scram bus voltage restored, and the main contacts reclosed before the auxiliary contacts opened. Measurements on the 12 contactors in the Safety System with an ohmmeter revealed that in 9 of the contactors the auxiliary contacts opened after the main contacts.

Notes

1. GEAP-13649, *Design and Testing of the Sefor Fast Reactivity Excursion Device (FRED)*, AEC Research and Development Report, January 1, 1970.

2. Richard E. Webb, "Some Autocatalytic Effects during Explosive Power Transients in Liquid Metal Cooled, Fast Breeder Nuclear Power Reactors (LMFBRs)" (PhD diss., Ohio State University, 1971).

3. David Okrent, *Nuclear Reactor Safety—On the History of the Regulatory Process*, University of Wisconsin Press, 1981. The same material, under the title *On the History of the Evolution of Light Water Reactor Safety in the United States*, appears on the internet at fissilematerials.org/library/OkrentReactorSafety.pdf.

4. *Southwest Experimental Fast Oxide Reactor.* Quarterly Plant Operations Report No. 6, August 1–October 31, 1970.

CHAPTER

7

A Picture is Worth a Thousand Words

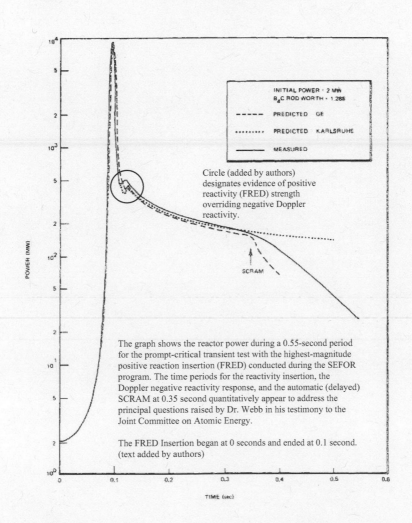

INITIAL POWER · 2 MW
B$_4$C ROD WORTH · 1.285

- - - - - PREDICTED GE

· · · · · · · · PREDICTED KARLSRUHE

————— MEASURED

POWER (MW)

Circle (added by authors) designates evidence of positive reactivity (FRED) strength overriding negative Doppler reactivity.

SCRAM

The graph shows the reactor power during a 0.55-second period for the prompt-critical transient test with the highest-magnitude positive reaction insertion (FRED) conducted during the SEFOR program. The time periods for the reactivity insertion, the Doppler negative reactivity response, and the automatic (delayed) SCRAM at 0.35 second quantitatively appear to address the principal questions raised by Dr. Webb in his testimony to the Joint Committee on Atomic Energy.

The FRED Insertion began at 0 seconds and ended at 0.1 second. (text added by authors)

TIME (sec)

CHAPTER
8 Conclusions

The "picture" (graph) shown in chapter 7 is remarkable. The measured reactor power rose from its starting value at the beginning of the test of 2 megawatts, as a result of the FRED reactivity insertion, to approximately 9,000 megawatts, before it was turned around and began decreasing as a result of the Doppler effect. Had the fuel been metallic rather than the oxide form, and if the fuel fissile concentration had been enriched to LMFBR levels, the Doppler effect would have been very much less than it was in the SEFOR (oxide) core, and core reactivity could have reached nuclear bomb damaging power. Instead, the Doppler effect stopped the acceleration of the power increase and caused it to immediately begin decreasing, which it continued to do until the reactor was SCRAMMED. The intentional SCRAM was purposely delayed for an additional period of 0.25 seconds (past the 0.1-second duration reactivity addition) in order to demonstrate the capability of the Doppler effect to turn around the power excursion.

Quoting from the General Electric report *SEFOR Core I Transients*,[1] "Eight super-prompt critical transients were performed in the Core 1 experimental program. The tests were initiated from nominal power levels of 2, 5, and 8 MW and resulted in peak power levels in excess of 9000 MW. . . . *The small secondary power peak on the 1.25$ transient occurs as additional positive reactivity insertion from the FRED **overrides** the initial Doppler feedback.* (emphasis added) The FRED rod reactivity insertion terminated at 120 msec, thus limiting the magnitude of the second peak."

This statement that the FRED positive reactivity insertion overrode the initial Doppler feedback appears to confirm Dr. Webb's warnings to the AEC that the ability of the Doppler effect to prevent a runaway power excursion was not unlimited. Indeed, it suggests that for the SEFOR reactor design, the upper limit of the Doppler capability was exceeded, if only slightly, in the highest positive reactivity insertions conducted at SEFOR.

Quoting from the General Electric report *SEFOR Experimental Results and Applications to LMFBR'S*,[2] "The SEFOR results aptly demonstrate the effectiveness of the Doppler effect in providing inherent stability and safety to LMFMR's. A particularly convincing demonstration was provided by the super-prompt transient tests in which poison rods worth up to 1.3$ in reactivity were ejected from the core from initial power levels up to 8 MW. In these tests the power quickly rose to ~10,000 MW as a result of the rapid reactivity of ~15 cents/second; however, even before all of the reactivity was inserted, *the Doppler effect stopped the rapid power rise and brought the reactor power to a level at which a delayed scram of ~350 msec could safely terminate the test.* (emphasis added) These results are in excellent agreement with predicted results based on pre-experimental analysis. . . . The good agreement between the SEFOR predicted and experimental results for both static power measurements and transient tests indicates that the Doppler theory as applied to LMFBR's is sound. The results of the various tests, therefore, may be used to provide a calibration of Doppler calculations for other reactor systems, taking care to use consistent data and analysis techniques throughout the analysis of the SEFOR results and the calculation of LMFBR power transients." *This statement further confirms that the inherent safety provided by the Doppler effect, however limited, is important because it allows time for the activation of the SCRAM mechanism to stop the runaway.*

In chapter 2, we made the following statement:

If, in Test No. 6, the increase in SEFOR's power level caused by the positive reactivity insertion had not hesitated as predicted by the Doppler effect; if the reactivity insertion had been accidentally maintained longer than the planned 0.1-second duration; and if the SCRAM procedure had accidentally failed—there was the real possibility that the power level might have increased to levels with the potential to rupture the containment.

A failure of the FRED system resulting in the poison rod suffering delayed reentry to the core beyond the planned 0.1-second duration was a risk factor that we cannot determine. In any case, we have no record that there was any such failure during any of the prompt-critical transient tests that we are considering. The possibility that the power increase would not hesitate as predicted was effectively made unlikely by the nearly two-year-duration experimental study of the reactor to determine, check, and double-check the calculation of the reactor core's Doppler Coefficient. The agreement between the measured and calculated performance of the reactor shown in the "picture" beginning this final chapter speaks powerfully to the expertise and care provided by the General Electric Company's design and experimental evaluation of the reactor before advancing to the conduct of the prompt-critical transient tests. Indeed, the agreement shown between all of the prompt-critical transient test measurements and the calculated values made before the experiments were conducted is conclusive evidence that the General Electric personnel conducting the tests had carefully examined the reactor's performance over a two-year period to verify and build confidence in the value of the Doppler coefficient in the two different SEFOR cores tested. *It also appears to indicate that GE demonstrated the maximum limits of the Doppler effect's capability to turn around a potential runaway explosion by intentionally bumping up against the limiting values they were confirming.*

If the reactivity experiments had for any reason exceeded the reactor's capability of turning the excursion around with the Doppler effect, the last resort would have been the SCRAM system. Our study of the extensive reports available to us indicated that the SCRAM system was well conceived and designed for its purpose, and it was extensively tested before the prompt-critical transient tests were undertaken. *However, we learned that SEFOR suffered at least one partial-SCRAM failure during its operation.* That failure did not occur during a prompt-critical transient test. We have not had the resources to calculate what the effect would have been if there had been a serious SCRAM failure during the prompt-critical transient tests such as the one depicted at the beginning of the chapter.

We do know the total hours the reactor was critical during its lifetime—3,895. If only one SCRAM failure had occurred during that operation, the failure frequency would be one in 3,895 hours or two per operating year. Such a frequency is not comforting considering the possible consequences.

If any of the accidental occurrences we have posited had occurred during the test considered in this final chapter, it seems highly probable that the SEFOR fuel temperature would have exceeded the melting temperature. Even partial liquefaction of the core could cause rearrangements of the enriched plutonium fuel, potentially resulting in powerful autocatalytic reactivity increases. If that had happened during the test "pictured," the chances that the resulting explosive overpressures would have ensured that the primary containment would not fail seem slim to us, as the SEFOR steel-reinforced concrete containment was designed to withstand an explosion of just 200 pounds TNT equivalent.

Notes

1. GEAP-13837, *SEFOR Core I Transients*, Breeder Reactor Department, General Electric Company, Sunnyvale, California, August 1972.

2. GEAP-13929, *SEFOR Experimental Results and Applications to LMFBR'S*, Breeder Reactor Department, General Electric Company, Sunnyvale, California, January 1973.

Postscript

Nuclear energy offers a Faustian bargain. It offers the World an inexhaustible source of energy. But in return, it demands a vigilance and longevity of our social sciences to which we are quite unaccustomed.

Alvin M. Weinberg, 1915–2006

The SEFOR project was conceived in the middle of the twentieth century during the period when the Liquid Metal Fast Breeder Reactor (LMFBR) was the AEC's highest priority. But the LMFBR program had strong opposition on economic grounds as well as concerns that fast reactors had the potential to suffer *nuclear* explosions that could not be economically contained to ensure against catastrophic releases of radioactive products, including the fissile fuel itself, in the form of aerosols to the environment. The purpose of the SEFOR program was to demonstrate the Doppler effect in an LMFBR fueled with plutonium oxide. That demonstration was successfully completed.

This book began to provide an accurate history of SEFOR. But our research quickly led us to the knowledge that the final experiments conducted at SEFOR for the Atomic Energy Commission prior to its closure in early 1972 involved intentional insertion of reactivity into the SEFOR core sufficient to change the nuclear reaction process from a state in which the chain reaction fission rate was controlled by slow neutrons with average velocities of about 4,500 miles per hour to a chain reaction state driven by "fast" neutrons with average velocities at least a thousand times greater. Such fast-neutron fission can very quickly drive the reaction at rates that are not controllable; this is nuclear fission explosion territory. But the AEC was confident that science had predicted a nuclear reaction effect, herein called the Doppler effect, that could result in a slowing of such "runaway" rates if the plutonium used as fuel was of the oxide form, PuO_2. As our country is now considering the expanded use of fast-fission reactors for electricity generation, driven powerfully by a motivation to provide an important solution to the looming climate change threat—we added an additional purpose: the provision of an accurate account of the risks that were taken in completing these experiments in the foothills of Arkansas's Ozark Mountains.

Also, we hoped to provide some closure to the questions of the residents in the neighborhood of the reactor. There is widespread fear of anything radioactive, and misconceptions about what occurred at SEFOR abound. At least for this group of people, though, we feel confident that this book provides information that will alleviate most of the concerns we are aware of. To complete a clean-slate assessment of the dangers that could potentially remain at the site, there are two principal questions about the disposition of the hazardous materials that were involved.

Sodium

In our review of the literature to determine where the sodium was disposed of shortly after the closure of SEFOR in early 1972, we found indications that ninety-some-odd barrels of sodium had been transported to the Nevada Low-Level Radioactive Waste Site near Beatty, Nevada. But our visit to the site failed to disclose any written records of verification.

Just before this book went to press in 2020, we discovered on the internet a report dated December 30, 2015, by the Nevada Department of Public Safety entitled: *Report on the October 18, 2015 Industrial Fire Incident at the Closed State of Nevada Low-Level Radioactive Waste Site.*[1]

Quoting from the Executive Summary of the report:

On October 18, 2015, an industrial fire incident occurred at the closed State of Nevada low-level radioactive waste disposal site located approximately 12 miles south of Beatty, Nye County, Nevada on US Highway 95 near milepost NY 48. . . .

The State of Nevada acquired this site in 1961 to receive low-level radioactive waste materials. These materials were buried for disposal at this site from 1962 until the site was closed in 1992. Materials were buried in numbered trenches and covered by an earth fill. This incident occurred at the east end of Trench 14 near the east perimeter of the closed waste disposal site.

Waste materials were buried in a variety of containers and packaging, including steel drums, cardboard boxes and wood crates. Over multiple decades of burial, the packaging materials have deteriorated and collapsed causing void spaces and the resulting settlement of the fill and cover material in several areas at the site.

Metallic sodium, packed in oil-filled steel drums, was received from at least three sources for burial at the east end of Trench 14 at this site. The sources included two (2) drums from a US Bureau of Mines Research Center in Boulder City, closed by that agency in the early 1970's; twenty-two (22) drums from Gulf-United Nuclear, Elmsford, New York, and ninety-two (92) drums from GE Nuclear Energy Division-SEFOR, Fayetteville, Arkansas.

Corrosion of the steel drums containing the metallic sodium over time allowed the packing fluid to drain out, leaving the metallic sodium exposed to the underground elements.

Approximately two weeks prior to the event, Desert Research Institute (DRI) instrumentation at the site reported 1.29 inches of rainfall on October 4 through 6, inclusive. On the day of the incident DRI instruments recorded an additional 0.57 inch of precipitation.

Although the original cover was designed and sloped to drain rainwater, there was evidence to indicate that portions of the cover were compromised due to settling and collapse of underlying waste containers and resulting subsidence and cracking of the cover, allowing the migration of rainwater into these areas.

The heavy precipitation prior to and on the day of the event saturated the earthen cover of the buried waste. Rainwater seeping through the compromised earth cover reached the metallic sodium causing an exothermic reaction between the water and the metallic sodium.

The reaction produced a large amount of heat and generated quantities of hydrogen gas. The volume of gas produced caused the eruption of the ground, expelling dirt, buried and corroded drums, and the products of the sodium-water reaction, primarily sodium hydroxide.

The heat generated by the sodium water reaction ignited combustible metals at the immediate site, resulting in a fire.

The fire continued to burn into the evening and early morning hours of the following day until all fuel had been consumed. At that point the fire extinguished itself.

The incident resulted in no injuries to personnel, the effects of the fire were contained to the immediate site, and there was no release of radioactive materials.

So much for the sodium.

Plutonium

This is a bit more uncertain, but we can be confident that dangerous amounts of plutonium do not remain near Strickler either. The brief documents describing the decommissioning of the SEFOR site when it was closed in 1972 indicated that the spent fuel from the reactor, which would have almost certainly included the plutonium remaining on the site, was transported by truck to Hanford.

We have identified only one other report that deals with the fate of the SEFOR plutonium, entitled *Processing of Non-PFP Plutonium Oxide in Hanford Plants*.[2] Quoting from that report:

The SEFOR campaigns at PUREX in December 1966 and April-May separated plutonium to produce plutonium nitrate solutions were shipped off site for use in production of SEFOR MOX fuel. There were some SEFOR returns of irradiated fuel as waste to the 200 Area burial grounds. The bulk of the SEFOR fuel reprocessing appears to have been conducted at Savannah River between 7/84 and 12/84. The fuel may have been sent to Savannah River rather than Hanford because the Hanford PUREX plant shut down at about the same time the SEFOR reactor was deactivated.

In short, aside from the account in the overview of Gerber,[3] quoted below, no written evidence for the processing of SEFOR MOX at Hanford was found in the technical literature.

In the later years of REDOX operation (1963–1967), Plutonium Recycle Test Reactor (PRTR) and Shippingport (Pennsylvania) Reactor fuels were processed. PUREX also reprocessed
some PRTR fuel in 1972, as well as some Southeast Experimental Fast Oxide Reactor (SEFOR) fuel. Core dissolving of
these mixed oxide fuels involved the use of a highly corrosive
mixture of nitric acid and hydrofluoric acid, with the dissolver
solution then blended with recycled uranium to achieve criticality control.

Instead, it seems that the supporting documents that described
reprocessing campaigns to recover plutonium from relatively high burn-
up Hanford fuel for use in preparing SEFOR MOX were misinterpreted
to draw the conclusion that the reprocessing of irradiated SEFOR fuel
occurred at Hanford. The dates of the processing reports (1963–1967)
alone indicate that reprocessing of irradiated SEFOR fuel did not occur
at Hanford because SEFOR did not go critical until April 1969. In any
case, the spent fuel, including the plutonium, is most likely at Hanford
or Savannah River, or further used for other purposes—it is no longer
near Strickler.

Final Thoughts

While the residents near the SEFOR site can be confident, we believe,
that the site no longer harbors dangers, there is a very important legacy of the SEFOR super-prompt-critical experiments that must not be
forgotten.

Today, as the site resumes an essentially greenfield condition, there
is a strong push by the current federal government to adopt fast-reactor
generation of electrical power in order to alleviate the climate change
problem that is resulting because of the increasing carbon dioxide concentration in our atmosphere.

We believe the important legacy of SEFOR is the knowledge
described in this book about the conduct at the site of super-prompt-
critical experiments. We acknowledge the apparent extreme caring
effort that seems to have been insisted upon by the General Electric
Company in the successful and safely conducted experiments that
demonstrated the important physics of the Doppler effect for the safe

operation of fast reactors fueled with mixed-oxide fuels containing plutonium and uranium.

But most importantly, we believe that there is a danger that the Doppler effect demonstration completed at SEFOR could leave the public with a false sense of security. While the Doppler effect is now demonstrated to be real and accurately predictable, we believe that any suggestion that it is any guarantee of safety regarding the possibility of nuclear explosions is not correct.

Instead, we believe that from the beginning of the SEFOR effort, there have existed real concerns by competent scientists and engineers that there are unpredictable effects that could attend rearrangement of the fissile material caused by accident or natural disasters that might override the Doppler effect.

In our judgment, this leaves us, at least presently, in the situation of being unable to confidently design a commercial fast reactor for electricity generation that will contain a worst-case accident explosion, ensuring against a catastrophic release of extremely dangerous radioactive materials to the environment. It seems to us that this predicament is foretold in the admonition provided at the beginning of this postscript by Alvin Weinberg.

Notes

1. *Report on the October 18, 2015 Industrial Fire Incident at the Closed State of Nevada Low-Level Radioactive Waste Site*, Nevada Department of Safety, Carson City, Nevada, December 30, 2015.

2. S. A. Jones and C. H. Delegard, *Processing of NON-PFP Plutonium Oxide in Hanford Plants*, PNNL-20246, WTP-RPT-211, March 2011.

3. M. S. Gerber, *The Plutonium Production Story at the Hanford Site; Processes and Facilities History*, WHC-MR-0521, Westinghouse Hanford Company, Richland, Washington, June 1996.

Index

When photographs and figures are referenced in this index, the entry is denoted with *p* or *f* respectively.